NATIONAL STANDARD
OF THE PEOPLE'S REPUBLIC OF CHINA

Unified Standard for Reliability Design of Engineering Structures

GB 50153 - 2008

Chief Development Department: Ministry of Housing and Urban-rural Development of the People's Republic of China

Approval Department: Ministry of Housing and Urban-Rural Development of the People's Republic of China

Implementation Date: July 1, 2009

China Architecture & Building Press

Beijing 2014

图书在版编目(CIP)数据

工程结构可靠性设计统一标准 GB 50153-2008/中华人民共和国住房和城乡建设部组织编译. —北京:中国建筑工业出版社,2014.9
(工程建设标准英文版)
ISBN 978-7-112-16942-9

Ⅰ.①工… Ⅱ.①中… Ⅲ.①工程结构-结构可靠性-设计规范-中国-英文 Ⅳ.①TU311.2-65

中国版本图书馆 CIP 数据核字(2014)第 119760 号

Chinese edition first published in the People's Republic of China in 2008
English edition first published in the People's Republic of China in 2014
by China Architecture & Building Press
No. 9 Sanlihe Road
Beijing, 100037
www.cabp.com.cn

Printed in China by BeiJing YanLinJiZhao printing CO., LTD

ISBN 978-7-112-16942-9(25738)

Announcement of Ministry of Housing and Urban-Rural Development of the People's Republic of China

No. 156

Announcement of Publishing the National Standard *Unified Standard for Reliability Design of Engineering Structures*

Unified Standard for Reliability Design of Engineering Structures has been approved as a national standard with a serial number of GB 50153－2008. It will be implemented on July 1, 2009. Therein, Articles 3.2.1 and 3.3.1 are compulsory provisions and must be enforced strictly. The original GB 50153－92 *Unified Standard for Reliability Design of Engineering Structures* shall be abolished simultaneously.

Authorized by Research Institute of Standards & Norms of the Ministry of Housing and Urban-Rural Development of the People's Republic of China, this Standard is published by China Architecture & Building Press.

Ministry of Housing and Urban-Rural Development of the People's Republic of China

November 12, 2008

Foreword

According to the requirements of Document Jian Biao [2003] No. 102 issued by the Ministry of Construction—"Notice on Printing the Development and Revision Plan of National Engineering Construction Standards in 2002 and 2003", China Academy of Building Research made a comprehensive revision for the national standard "Unified Standard for Reliability Design of Engineering Structures" GB 50153—92 jointly with the departments concerned.

This standard was revised by borrowing actively from the international standard ISO 2394: 1998 "General Principles for Structural Reliability" issued by International Organization for Standardization (ISO) and the European standard EN 1990: 2002 "Bases of Structural Design" approved by European Committee for Standardization (CEN), carrying out the principle of preceding from the actual conditions of China carefully, summing up the practical experience of China's large-scale engineering and carrying through the guiding principle of sustainable development. There have been some significant extensions in this revised new standard over its previous edition. This revised new standard specifies the basic contents on the design basis of engineering structures and is a basic standard for design of engineering structures.

This revised new standard uniformly specifies the fundamentals, basic requirements and basic methods for the design of engineering structures in various civil engineering fields such as architectural engineering, railway engineering, highway engineering, harbor and harbor engineering and water conservancy hydroelectric engineering, in order to make these = engineering fields be provided with consistency and coordination on processing the structural reliability problem and link them up to the world. General requirements for the design of engineering structures in various civil engineering fields are included in the text of this standard, while the specific requirements for special branches and problems are included in the Appendixes. The main contents of this standard are as follows: General Provisions, Terms and Symbols, Basic Provisions, Limit State Design Principles, Actions and Environmental Influence on the Structure, Material and Geotechnical Properties/Geometrical Parameters, Structural Analysis and Test- assisted Design and Partial Factor Design Methods.

The bold provisions in this standard are compulsory provisions and must be enforced strictly.

Ministry of Housing and Urban-rural Development is in charge of the administration of this standard and the explanation of compulsory provisions. And China Academy of Building Research is responsible for the explanation for the specific technical contents. In order to improve the standard quality, all relevant organizations are kindly requested to sum up and accumulate - experience in practices during the process of implementation of this standard. The relevant opinions and advice, whenever necessary, may be posted or passed on to China Academy of Building Research (Address: No. 30, North Third Ring East Road, Beijing, 100013, China) for future reference.

Chief Development Organization of this standard:

China Academy of Building Research

Participating Development Organizations of this standard:

China Academy of Railway Sciences

The Third Railway Survey and Design Institute Group Co. ,Ltd.

CCCC Highway Consultants Co. ,Ltd.

China Communications Water Transportation Planning and Design Institute Co. ,Ltd.

Hydropower and Water Resources Planning and Design General Institute, Ministry of Electric Power

Water Resources and Hydropower Planning and Design General Institute, Ministry of Water Resources

Dalian University of Technology

Xi'an University of Architecture and Technology

Shanghai Jiao Tong University

China Association for Engineering Construction Standardization

Chief Drafting Staff of this standard:

Yuan Zhenlong　Shi Zhihua　Li Mingshun　Hu Dexin

Chen Jifa　Li Yungui　Di Xiaotan　Liu Xiaoguang

Li Tiefu　Zhang Yuling　Zhao Junli　Du Tingrui

Yang Songquan　Shen Yisheng　Zhou Jianping　Lei Xingshun

Gong Jinxin　Yao Jitao　Bao Weigang　Yao Mingchu

Liu Xila　Shao Zhuomin　Zhao Guofan

Contents

1 General Provisions

1.0.1 This standard is established with a view to unifying the fundamentals, basic requirements and basic methods for design of various engineering structures such as buildings, railways, highways, port and harbor structures as well as water conservancy and hydropower, and making the structures meet the sustainable development requirements as well as requirements on safety, economy, state-of-art technology and high quality.

1.0.2 This standard is applicable to design of the members of the whole structures, their components and the foundations; to the design of structures for construction- and service periods; and to the reliability assessment of the structures.

1.0.3 Probability-based limit state and partial factor format should be adopted for the design of engineering structures. The design of structures may be carried out according to engineering experience or necessary experimental investigation if statistical data is insufficient. In some cases, such empirical methods as allowable stress or slumped safety factor may also be used.

1.0.4 Standards for design of engineering structures and other relevant standards shall comply with the ground rules specified in this standard and establish corresponding specific specifications.

1.0.5 The design of engineering structures shall not only comply with the specifications specified in this standard, but those in the current relevant national ones.

2 Terms and Symbols

2.1 Terms

2.1.1 Structure

A system which is composed of various connecting parts and rigid to withstand the actions.

2.1.2 Structural Member

Part that may be separated physically from the structures.

2.1.3 Structural System

All the load-bearing members and the way in which these members function together.

2.1.4 Structural Model

The idealized structural system used for structural analysis and design.

2.1.5 Design Working Life

The intended period specified in the design, during which the structure and its members may be used as intended purpose with anticipated maintenance but without major repair being necessary.

2.1.6 Design Situations

A group of design conditions which represent the practical situations in a certain period of time. The structures shall not exceed the related limit state in design under this group of conditions.

2.1.7 Persistent Design Situation

Design situation that is relevant during a period of the same order as the design working life of the structure.

2.1.8 Transient Design Situation

Design situation that is relevant during a period much shorter than the design working life of the structure and which has a high probability of occurrence.

2.1.9 Accidental Design Situation

Design situation involving exceptional conditions of the structure or its exposure.

2.1.10 Seismic Design Situation

The design situation when the structures are exposed in an earthquake.

2.1.11 Load Arrangement

The reasonable determination for the positions, magnitude and orientation of the free action in structural design.

2.1.12 Load Case

The compatible load arrangementof the fixed variable actions, permanent actions and free actions as well as deformation and geometrical deviation, which are considered simultaneously for specific verification purposes.

2.1.13 Limit States

A certain specific state that if it is exceeded which some functional requirement of the entire structures or a part of the structures specified in the design may not be fulfilled.

2.1.14 Ultimate Limit States

The state in which the structures or structural members reach their maximum load carrying capacity or the deformation that is not suitable for sustaining load.

2.1.15 Serviceability Limit States

The state in which the structures or structural members reach a certain specific serviceability or durability limit.

2.1.16 Irreversible Serviceability Limit States

Serviceability limit states that may be exceeded under an action and is irrecoverable after this action is eliminated.

2.1.17 Reversible Serviceability Limit States

Serviceability limit states that may be exceeded under an action and is recoverable after this action is eliminated.

2.1.18 Resistance

The capacity of structures or structural members withstand the effect of actions.

2.1.19 Structural Integrity (Structural Robustness)

The capacity of structures keep stable integrity and be absent of consequences of failure due to disproportionate causes to the origin such as fire, explosion, impact or human error and other accident events .

2.1.20 Progressive Collapse

The initial local failures, spread from members to members, result eventually in catastrophic collapse of the whole structures or disproportionate local damage to the origin causes.

2.1.21 Reliability

The capability of structures fulfill the prescribed functions within the intended period under specified conditions.

2.1.22 Degree of Reliability (Reliability)

The probability content of the structures fulfill the prescribed functions within the intended period under specified conditions.

2.1.23 Probability of Failure p_f

The probability of structures that may not fulfill the prescribed function.

2.1.24 Reliability Index β

The numerical value used to measure the structural reliability. The relation between the reliability index β and the probability of failure p_f is $\beta = -\Phi^{-1}(p_f)$. Where, $-\Phi^{-1}(\cdot)$ is the inverse function of the standard normal distribution function.

2.1.25 Basic Variable

A group of variables used to represent the physical quantities, such as action and environmental influence, material and geotechnical properties as well as geometrical parameters.

2.1.26 Performance Function

A function of basic variables and used to characterize a kind of structural function.

2.1.27 Probability Distribution

A statistical description of random variable and expressed in probability density function or probability distribution function.

2.1.28 Statistical Parameter

Parameters for describing the statistical characteristics of random variable such as the mean, standard derivation and so on.

2.1.29 Fractile

The value corresponding to a certain probability content in the probability distribution function of the random variable.

2.1.30 Nominal Value

The value determined by non-statistical method.

2.1.31 Limit State Method

The design method used to make the structures not exceeding certain specified limit states.

2.1.32 Permissible (Allowable) Stress Method

The design method used to make the stress of structure or foundation under the characteristic value of actions not exceeding the specified permissible stress (material or geotechnical characteristic value divided by a certain lumped safety factor).

2.1.33 Lumped Safety Factor Method

The design method used to make the ratio of the resistance characteristic value to the action characteristic value not being lower than a certain specified lumped safety factor.

2.1.34 Action

The concentrated force or distributed force (direct action, also known as load) applied on the structures or cause of structural imposed deformation or restrained deformation on the structure (indirect action).

2.1.35 Effect of Action

The reaction of the structures or structural members caused by actions.

2.1.36 Single Action

The action which is statistically independent from any other action on the structures in both time and space.

2.1.37 Permanent Action

The action which is always existed in the design working life and the variation in magnitude may be negligible comparing with its average value or is monotonic tend to a certain limit.

2.1.38 Variable Action

The action whose magnitude varies with time and the variation in magnitude may not be negligible comparing with the average value in the design working life.

2.1.39 Accidental Action

The action which may not appear in the design working life, or is significant in magnitude and short in duration once it appears.

2.1.40 Seismic Action

The action generated by earthquake on the structures.

2.1.41 Geotechnical Action

The action which is transferred to the structures by ground, fill or underground water.

2.1.42 Fixed Action

The action provided with fixed space distribution on the structure. Its influence on the whole structures may be determined once its magnitude and orientation in a point of structure are determined .

2.1.43 Free Action

The action with random space distribution in the prescribed limit on the structure.

2.1.44 Static Action

The action that the generated acceleration on the structure is negligible.

2.1.45 Dynamic Action

The action that the generated acceleration on the structures is not negligible.

2.1.46 Bounded Action

The action that may not be exceeded and its bound may be exactly or approximately determinate.

2.1.47 Unbounded Action

The action with no definite bound.

2.1.48 Characteristic Value of an Action

The main representative value of an action which may be determined according to the statistic of the observation data, natural limit of the action or engineering experience.

2.1.49 Design Reference Period

The time parameter for assessing the values of variable actions, etc..

2.1.50 Combination Value of a Variable Action

The action value that makes the exceeding probability of the combined effect of action tend to be uniform with that of the characteristic value effect of action when it appears solely; or action value makes the structures have the specified reliability index for combination. It may be expressed by multiplying the combination value factor ($\psi_c \leqslant 1$) on the characteristic value of an action.

2.1.51 Frequent Value of a Variable Action

The action value that the total time being exceeded accounts for a fraction of the design reference period; or the action value that the frequency being exceeded is limited to the assigned frequency. It may be expressed by multiplying the frequent value factor ($\psi_f \leqslant 1$) on the characteristic value of an action.

2.1.52 Quasi-permanent Value of a Variable Action

The action value that the total time being exceeded accounts for a large part of the design reference period. It may be expressed by multiplying the quasi-permanent value factor ($\psi_q \leqslant 1$) on the characteristic value of an action.

2.1.53 Accompanying Value of a Variable Action

The value of a variable action that appears with the leading action in the considering combinations. It may be either combination value, frequent value or quasi-permanent value of an action.

2.1.54 Representative Value of an Action

The action value used in limit state design. It may either be characteristic value of an action or accompanying value of a variable action.

2.1.55 Design Value of an Action

The product of representative value of an action by its partial factor.

2.1.56 Combination of Actions (Load Combination)

A group of design values used to verify the structural reliability of a certain limit state under the simultaneous influence of different actions.

2.1.57 Environmental Influence

The mechanical, physical, chemical or biologic adverse effects of the environment on the structures. The environmental influence may lead to deterioration of the structural material property, reduce the safety, serviceability and durability of the structures.

2.1.58 Characteristic Value of a Material Property

A certain fractile of the probability distribution or the nominal value of a material property which meets the specified quality .

2.1.59 Design Value of a Material Property

The value obtained by the characteristic value of a material property divided by the partial factor of a material property.

2.1.60 Characteristic Value of a Geometrical Parameter

The nominal value of a geometrical parameter specified in design or a certain fractile of the probability distribution of a geometrical parameter.

2.1.61 Design Value of a Geometrical Parameter

The value obtained by the characteristic value of a geometrical parameter adding or subtracting an additional quantity of a geometrical parameter.

2.1.62 Structural Analysis

The process determining the effect of actions on the structures.

2.1.63 First Order Linear-elastic Analysis

The structural analysis conducted for the geometrical body of the initial structures using elastic theory based on the linear stress-strain or bending moment-curvature relations.

2.1.64 Second Order Linear-elastic Analysis

The structural analysis conducted for the geometrical body of the deformed structures using elastic theory based on the linear stress-strain or bending moment-curvature relations.

2.1.65 First Order (or Second Order) Linear-elastic Analysis with Redistribution

The first or second order structural analysis with adjustment to the internal force which is equilibrium to the external action, and no explicit rotation verification is needed.

2.1.66 First Order Non-linear Analysis

The structural analysis conducted for the geometrical body of the initial structures based on the nonlinear material constitutive model.

2.1.67 Second Order Non-linear Analysis

The structural analysis conducted for the geometrical body of the deformed structures based on the nonlinear material constitutive model .

2.1.68 Elasto-plastic Analysis (First or Second Order)

The structural analysis conducted based on bending moment-curvature relation consisting of linear-elastic branch and subsequent non-hardening branch.

2.1.69 Rigid Plastic Analysis

The structural analysis conducted directly to determine the ultimate load carrying capacity of the geometrical body of the initial structures by ultimate analysis theory, with assumed elastic-perfect plastic bending moment-curvature relation.

2.1.70 Existing Structure

Various engineering structures available and in service.

2.1.71 Assessed Working Life

Working life of the existing structure evaluated by reliability assessment under specified conditions.

2.1.72 Load Testing

The test carried out to evaluate the performance of the structures or structural members or to predict their load carrying capacity through exerting load directly on the structure.

2.2 Symbols

2.2.1 Symbols for Latin capitals:

A_{Ek}——Characteristic value of the seismic action;

A_d——Design value of the accidental action;

C——Limit specified for deformation and crack in serviceability limit state verification;

F_d——Design value of an action;

F_r——Representative value of an action;

G_k——Characteristic value of a permanent action;

P——Relevant representative value of a prestress action;

Q_k——Characteristic value of a variable action;

R——Resistance of structures or structural members;

R_d——Design value of the resistance of structures or structural members;

S——Effect of action of structures or structural members;

$S_{A_{Ek}}$——Effect of the characteristic value of a seismic action;

S_{A_d}——Effect of the design value of an accidental action;

S_d——Design value of effect of the combination of actions;

$S_{d,dst}$——Design value of effect of an destabilishing action;

$S_{d,stb}$——Design value of effect of a stabilishing action;

S_{G_k}——Effect of the characteristic value of a permanent action;

S_P——Effect of the relevant representative value of a prestress action;

S_{Q_k}——Effect of the characteristic value of a variable action;

T——Design reference period;

X——Basic variable.

2.2.2 Symbols for Latin smalls:

a——Geometric parameter;

a_d——Design value of a geometrical parameter;

a_k——Characteristic value of a geometrical parameter;

f_d——Design value of a material property;

f_k——Characteristic value of a material property;

p_f——Calculation value of probability of failure of a structural member.

2.2.3 Symbol for U. C. Greek:

Δ_a——Additional quantity of a geometrical parameter.

2.2.4 Symbols for L. C. Greeks:

β——Reliability index of a structural member;

γ_0——Importance Factor of a structure;

γ_I—— Importance Factor for a seismic action;

γ_F——Partial factor of an action;

γ_G——Partial factor of a permanent action;

γ_L——Load adjustment factor taking account of the design working life of a structure;

γ_M——Partial factor of a material property;

γ_Q——Partial factor of a variable action;

γ_P——Partial factor of a prestress action;

ψ_c——Combination value factor of an action;

ψ_f——Frequent value factor of an action;

ψ_q——Quasi-permanent value factor of an action.

3 Basic Provisions

3.1 Basic Requirements

3.1.1 The design, construction and maintenance of structures shall make the structure meet the specified various functional requirements in appropriate reliability and an economical way within the specified design working life.

3.1.2 Structures shall meet the following functional requirements:

1 Be able to withstand various actions that may present in the construction and service periods;

2 Keep excellent serviceability;

3 Possess adequate durability;

4 Be able to maintain adequate load-carrying capacity within the stipulated time in case of fire hazards;

5 Be able to maintain necessary integral stability and be absent of consequences of failure that disproportionate to the original causes such as fire, explosion, impact or human error and other accidents in order to prevent the structures from progressive collapse.

Notes: 1 All necessary steps shall be taken to prevent the structures from progressive collapse for important structures; appropriate measures should be adoptedfor ordinary structures.

2 "Impact" refers to abnormal impact for port and harbor structures.

3.1.3 Appropriate measures shall be taken according to the following requirements in the structural design in order to ensure there are no or less possible damages:

1 Avoid, eliminate or reduce the hazards risk that may be encountered by the structures;

2 Select structural types which are insensitive to the possible hazards;

3 Select structural types, in which the remained structures are able to be survived when an individual member or finite parts of the structures are removed accidentally or the structures are subjected to a receivable local damage;

4 Structural systems with no destruction warning should not be adopted;

5 To be robust in accident situation.

3.1.4 The following measures should be taken to meet the basic requirements for structures:

1 Select appropriate materials;

2 Select rational design scheme and detailing;

3 Establish appropriate control measures for design, manufacture, construction and utilization of the structures.

3.2 Safety Classes and Reliability

3.2.1 Different safety classes shall be adopted in the design of engineering structures according to the consequences (such as imperil the life of a person, create economic loss, and influence the society or environment) that possibly occur for the structural damages. The division for safety classes of the engineering structures shall be in accordance with those specified in Table 3.2.1.

Table 3.2.1 Safety Classes of Engineering Structures

Safety Classes	Consequence of failure
Class Ⅰ Class Ⅱ Class Ⅲ	Very serious Serious Less serious

Note: The safety class shall be class Ⅰ, Ⅱ and Ⅲ for important, ordinary and secondary structures respectively.

3.2.2 The safety classes of various structural members of the structures should be identical with that of the structures, while parts of the structural members may be adjusted, but shall not be less than class Ⅲ.

3.2.3 The setting of reliability of the structural members shall be determined according to the safety classes, failure modes and economic factors. Different reliabilities may be set for the safety and serviceability of the structures.

3.2.4 The reliability of a structural member should be descripted in term of reliability index β if there is sufficient statistical data. The reliability index in the design of structural members may be determined according to the reliability calibration for the existing structural members combined with the experience and economic factors.

3.2.5 The deference of reliability indexes between immediate different safety classes of structural members should be 0.5.

3.3 Design Working Life and Durability

3.3.1 The design working life of structures shall be specified in the design of engineering structures.

3.3.2 The design working life of building structures, railway bridge and culvert structures, highway bridge and culvert structures and port and harbor structures shall be in accordance with those specified in Appendix A.

Notes: 1 The design working life of other engineering structures shall be in accordance with the relevant regulations specified in the current national standards;

2 The design working life of special engineering structures may be otherwise specified.

3.3.3 The environmental influence shall be evaluated in the design of engineering structures. If the environment has a large impact on the durability of the structures , the corresponding structural materials, detailing, protective measures and construction quality requirements shall be adopted according to different environmental classifications. In addition, periodic repair and maintenance systems shall be established during the service period in order to the safety or serviceability of structures not being affected due to the material deterioration within the design working life.

3.3.4 The influence of environment on the durability of the structures may be evaluated by engineering experience, experimental study, calculation or comprehensive analysis.

3.3.5 The division of environmental classifications and the corresponding design/ construction / service/maintenance requirements shall be in accordance with those specified in the current relevant national standards.

3.4 Reliability Management

3.4.1 To guarantee the engineering structures are provided with specified reliability, controls shall be carried out for the material property, construction quality, operation and maintenance of the structures except for the necessary design calculation. The specific measures shall be in accordance with Appendix B and those specified in reconnaissance/design/construction/maintenance-related standards.

3.4.2 The design of engineering structures must be conducted by technicians with corresponding qualification.

3.4.3 The design of engineering structures shall be in accordance with the relevant load/-seismic/foundation requirements in the current national standards as well as the requirements specified in various material structural design codes.

3.4.4 Necessary protective measures shall be taken for any possible accidental action and environmental influence that the structures may encounter in the design of engineering structures.

3.4.5 Quality control shall be carried out for the adopted materials in construction and manufacturing of the engineering structures-. Final acceptance shall be conducted according to those specified in the current relevant national standards.

3.4.6 Engineering structures shall be used according to the specified purposes. And periodic inspection for the structural condition and necessary maintenance also shall be conducted. When the purposes of utilization need to be changed, recheck for the design shall be made and necessary safety measures shall be adopted.

4 Principles For Limit State Design

4.1 Limit States

4.1.1 The limit states includes ultimate limit states and serviceability limit states. Both the two kind of limit states shall meet the following requirements:

1 Ultimate limit states

The ultimate limit state are considered to be exceeded if the structure or structural member is subjected to one of the following conditions:

1) Structural members or its connection are failed due to the strength of materials being exceeded or they are unsuitable for continuing to load due to excessive deformation;
2) The whole structures or parts of them lose equilibrium as a rigid body;
3) The structure become an alterable system;
4) The structure or structural member loses its stability;
5) Progressive collapse occurs for the structure due to local failure;
6) The structure fails due to its foundation loses bearing capacity;
7) Fatigue failure of the structure or structural member.

2 Serviceability limit states

The serviceability limit states are considered to be exceeded if the structure or structural member is subjected to one of the following conditions:

1) Deformation influencing the serviceability or appearance;
2) Local damage influencing the serviceability or durability;
3) Vibration influencing the serviceability;
4) Other specific states influencing the serviceability.

4.1.2 Definite signs or limits shall be specified for various limit states of the structures.

4.1.3 The different limit states of the structures shall be calculated and verified respectively in the structural design. The calculation or verification may be conducted only for the critical limit state in the considering limit states.

4.2 Design Situations

4.2.1 The following design situations shall be distinguished for the design of engineering structures:

1 Persistent design situation is applied to a normal condition of structure;

2 Transient design situation is applied to a temporary condition appeared to the structure, including the situations of construction and maintenance;

3 Accidental design situation is applied to an abnormal condition happened to the structure, including fire, explosion or impact;

4 Seismic design situation is applied to structures subjected to an earthquake. Seismic design situation must be taken into account for the structures in the earthquake-prone area.

4.2.2 Different design situations shall be adopted according to structural system, reliability level, basic variables and combination of actions in the design of engineering structures .

4.3 Limit State Design

4.3.1 The following limit state designs shall be carried out respectively for the four kinds of design situations of structures specified in Article 4.2.1:

1 Ultimate limit state design shall be carried out for the four design situations;

2 Serviceability limit states design also shall be carried out for the persistent design situation;

3 Serviceability limit states design may be carried out as required for the transient design situation and seismic design situation;

4 Serviceability limit states design may not be carried out for the accidental design situation.

4.3.2 The following combinations of actions shall be adopted according to different design situations when the ultimate limit states design is carried out:

1 Basic combination, which is used for persistent design situation or transient design situation;

2 Accidental combination, which is used for accidental design situation;

3 Seismic combination, which is used for seismic design situation.

4.3.3 The following combinations of actions may be adopted if the serviceability limit states design is carried out:

1 Characteristic combination, which should be used for irreversible serviceability limit states design;

2 Frequent combination, which should be used for irreversible serviceability limit states design;

3 Quasi-permanent combination, which should be used for serviceability limit states design whose long-term effect is the decisive factor.

4.3.4 The design of engineering structures shall be carried out by adopting its most unfavorable design value for each combination of actions.

4.3.5 The limit states of structures may be described by the following limit state equation:

$$g(X_1, X_2, \cdots, X_n) = 0 \tag{4.3.5}$$

Where $g(\cdot)$——performance function of the structure;

$X_i (i = 1, 2, \cdots, n)$——basic variables, which refer to various actions and environmental influences on the structures, materials and geotechnical properties, as well as the geometrical parameters. The basic variable shall be taken as random variable in the reliability analysis.

4.3.6 The limit state design of structure shall meet the following requirements:

$$g(X_1, X_2, \cdots, X_n) \geqslant 0 \tag{4.3.6-1}$$

As a simple case of the effect of action and resistance of the structure being integrated variables, the limit state design for the structure shall meet the following requirement:

$$R - S \geqslant 0 \tag{4.3.6-2}$$

Where R——resistance of the structure;

S——effect of action on the structure.

4.3.7 The design of structural members shall meet the requirements specified in Article 4.3.6 of this Chapter with specified reliability.

4.3.8 The structural members should be designed according to the specified reliability index, by means of representative values of actions, characteristic values of a material properties, characteristic values of geometrical parameters and limit state design expressions with corresponding partial factors. Reliability index-based method specified in Appendix E may also be used directly if it is conditional.

5 Actions on the Structure and Environmental Influence

5.1 General Requirements

5.1.1 All the possible actions ,including direct actions and indirect actions, on the structure and the environmental influence shall be considered in the design of engineering structures.

5.2 Actions on the Structure

5.2.1 Each action may be respectively considered to be a single action on the structure if they are mutually independent in time and space. . They may be considered as a single action if they closely correlate to each other and are possibly to occur with their maximum values.

5.2.2 The joint influence of single actions applied on the structure shall be taken into account through combination of actions (load combination). No combination is required for actions which impossibly occur at the same time.

5.2.3 Actions on the structure may be classified according to the following properties:

1 According to variation with time:

1) Permanent action;

2) Variable action;

3) Accidental action.

2 According to variation in space:

1) Fixed action;

2) Free action.

3 According to reaction characteristics of the structure:

1) Static action;

2) Dynamic action.

4 According to the fact that whether there is limit or not:

1) Bounded action;

2) Unbounded action.

5 Other classifications.

5.2.4 The time history of actions on the structure should be described by stochastic process model . However, different methods may be adopted for simplification.

Random variable model may be adopted for permanent action in the reliability design of the structure.

Simplified stochastic process model may be adopted for variable action in the combination of actions. The determination of the representative value of a variable actioncan based on the probability model of the random variable of maximum value of variable action in the design reference period.

5.2.5 The statistical parameters and probability distribution of permanent and variable actions shall be determined based on the observation data and tested using the hypothesis testing method of the parameter estimations and probability distribution if they are adopted as the random variables.

The testing significance level may be 0.05.

5.2.6 The characteristic value of an action shall be determined according to some unfavorable statistical characteristics of the probability distribution in the design reference period if there is sufficient data. , The probabilistic definition of this statistical characteristic may be on the same basis if it is possible. , The characteristic value of an action may be determined according to the engineering experience if data is not available. For bounded action with definite bounded value, the characteristic value of an action shall take its bounded value.

Note: Characteristic value of a variable action may be determined according to the principles specified in Appendix C.

5.2.7 Different representative values of an action shall be adopted for actions which may present at the same time in the corresponding combination of actions when the engineering structures are designed according to different limit states. The representative values of variable action include characteristic value, combination value, frequent value and quasi-permanent value, and they may be expressed through the characteristic value of a variable action respectively multiplying by combination value factor ψ_c, frequent value factor ψ_f and quasi-permanent value factor ψ_q, all of which are not larger than 1.

Note: Combination value, frequent value and quasi-permanent value of a variable action may be determined according to the principles specified in Appendix C.

5.2.8 The design value shall be adopted in design for accidental action. The design value of the accidental action shall be determined according to specific conditions and its possible maximum value or specific requirements in the relevant standards.

5.2.9 The characteristic value shall be adopted in design for seismic action. It shall be determined according to the return period of the seismic action which should adopt 475 years or other values according to specific conditions.

5.2.10 Appropriate mathematical model may be established to characterize the magnitude, position, direction and duration of the action of different source and mechanism if the actions on the structure are quite complex and may not be described directly.

A general mathematical model may be adopted for the magnitude, F, of actions on the structure, as follows:

$$F = \varphi(F_0, \omega) \tag{5.2.10}$$

Where $\varphi(\cdot)$ ——a function;

F_0——basic action, which may be time-dependent and space-dependent (random or nonrandom) and generally structure-independent ;

ω——random or nonrandom variable which is used to convert F_0 into F and related to structural properties.

5.2.11 The structure shall be described by dynamic model if the dynamic property of the structure is significance. Dynamic analysis of the structure shall be based on its rigidity, damping and inertia of mass of all parts on the structure. When simplified analysis is allowed, "quasi-static action" response may be calculated as the response of the dynamic action by multiplying a dynamic coefficient.

5.2.12 Possible load arrangements shall be taken into account and combined with fixed action as a load case for the free action to verify a certain specific limit state of the structure.

5.3 Environmental Influence

5.3.1 Environmental influence include permanent influence, variable influence and accidental influence.

5.3.2 The environmental influence on the structure shall be described quantitatively. If it is impossible, qualitative description may be conducted through grading. In addition, corresponding measures shall be taken in the design.

6 Material and Geotechnical Properties and Geometrical Parameters

6.1 Materials and Geotechnical Properties

6.1.1 The physical and mechanical properties of the material as well as geotechnical properties such as strength, elastic modulus, deformation modulus, constrained modulus, internal friction angle and cohesion shall be determined according to the related test method.

6.1.2 The material property should be described by random-variable probability model. The determination of statistical parameters and probability distribution of the material property shall be based on the test data and hypothesis testing. The testing significance level may be 0.05.

6.1.3 When the actual material property of the structure is determined based on the test results of the standard specimens , the difference between the actual structure and the standard test specimen, and the actual operating conditions and the standard test conditions shall be taken into account. The relation between the material property of the structure and that of the standard test specimen shall be described by conversion coefficient or function established based on comparative test or be determined via engineering judgment. The uncertainty of the material property of the structure shall be composed of two parts, namely, the uncertainty of the material property of the standard specimen and the uncertainty of the conversion coefficient or function.

The geotechnical property and bearing capacity of the foundation and pile foundation shall be determined by direct or indirect methods, such as in-situ test and indoor test. And the influence of such factors as the difference between the indoor/outdoor conditions and the actual structure conditions due to the disturbance of the sampling, as well as the deviation of the calculation also shall be taken into account.

6.1.4 The probability distribution of the material strengths should adopt normal or logarithmic normal distribution.

The characteristic value of mate rial strength s may be determined according to 0.05 fractile of probability distribution. And the characteristic values of the elastic modulus, poisson ratio and other physical properties of the materials may be determined according to 0.5 fractile of probability distribution.

The characteristic value of material property may be specified in the relevant standards or be determined via engineering judgment if the test data is insufficient .

6.1.5 The characteristic value of geotechnical property should be determined according to the results of the in-situ test and indoor test and met the requirements in the relevant standards.

The characteristic value of geotechnical property may be determined according to a certain fractile of its probability distribution if it is conditional.

6.2 Geometrical Parameters

6.2.1 The geometrical parameter, a, of the structure or structural member should be described by a random variable model, and its statistical parameters and probability distribution shall be based on the test data under the normal production condition and be determined by hypothesis testing method in statistics.

The statistical parameter of geometrical parameter may be determined according to the tolerance specified in the relevant standards if the test data is insufficient.

The geometrical parameter may be taken as the deterministic variable if its variability has a little effect on the resistance and other properties of the structure.

6.2.2 The characteristic value of a geometrical parameter may be the nominal value in design, or be determined according to a certain fractile of the probability distribution.

7 Structural Analysis and Test-assisted Design

7.1 General Requirements

7.1.1 Structural analysis may be conducted by calculation, model test or prototype test.

7.1.2 The precision of structural analysis shall meet the structural design requirements. Experimental verification should be conducted if necessary,.

7.1.3 The influence of environment on the material properties and member/ structure performance should be taken into account in structural analysis.

7.2 Structural Model

7.2.1 The fundamental assumption and computation model adopted for the structural analysis shall be able to reasonably describe the structural response under the considered limit state.

7.2.2 One-dimensional, two-dimensional or three-dimensional model may be adopted in structural analysis according to the physical circumstances of the structure.

7.2.3 Simplified or approximate assumptions adopted in the structural analysis shall be on the theoretical or test basis, or be verified through engineering verification.

7.2.4 The structural deformation shall be taken into account in the structural analysis if the influence of the action increase with the increase of deformation of the structure obviously.

7.2.5 The uncertainty of the computation model of the structure shall be considered by one or several additional basic variables in the limit state equation. The probability distribution and statistical parameters of the additional basic variables may be determined via statistical analysis according to the difference between the calculation results of the adopted model and the calculation results by exact method or the actual observation, or determined via engineering judgment.

7.3 Action Model

7.3.1 Only the maximum value and minimum value of an action in the design reference period need to be considered for static analysis which is time-independent or is exclusive of the cumulative effect. More description shall be focused on the detailed process if the dynamic performance is dominant.

7.3.2 The upper and lower limit of action parameters shall be provided and compared in order to determine the adverse effect if it cannot be determined accurately.

7.3.3 The worst load arrangement for free action on the structure shall be determined according to its possible spatial position, magnitude and orientation.

7.3.4 The geotechnical action may be modeled by appropriate equivalent spring or damper if the interaction between the structure and the foundation is taken into account.

7.3.5 The dynamic effect may be considered by incorporating the dynamic action response into the static action or by multiplying the equivalent dynamic amplification factor on the static action if the dynamic action may be treated as a pseudo-static action.

7.3.6 Serviceability limit states of the structures shall be verified if it may be exceeded resulting from the amplitude, speed and acceleration of the dynamic action.

7.4 Analysis Method

7.4.1 The structural analysis shall be conducted by linear/nonlinear analysis or test method according to structural types, material properties and features of service and so on. The elastic theory may be used if the structure works always in the elastic stage. Otherwise, the plasto-elasticity theory should be used.

7.4.2 Plastic theory may be used in the structural analysis if the structure subjected to static or less cycles repeated action and sufficient plastic deformation may be generated before it reaches the limit state. The plastic theory shall be prohibited to be used if the load carrying capacity of the structure is dominate by brittle failure or stability.

7.4.3 The dynamic analysis shall be carried out if larger acceleration may be generated on the structure by dynamic action.

7.5 Test-assisted Design

7.5.1 Test-assisted design may be carried out in the cases of no appropriate analysis models may be used. The methods should be in accordance with those specified in Appendix D.

7.5.2 The same reliability level with the relevant design situations shall be achieved if the test-assisted design is adopted for the structure, and the influence of test sample size on the statistical uncertainty of the relevant parameters shall be taken into account.

8 Partial Factor Design Methods

8.1 General Requirements

8.1.1 The partial factors in the limit state design expressions of the structural member should be determined based on reliability analysis and specified reliability index combined with engineering judgment.

The partial factors may be specified in the relevant standards according to traditional or empirical design methods if no statistical data available.

8.1.2 The design value of a basic variable may be determined according to the following requirements:

1 The design value of an action, F_d, may be determined as follows:

$$F_d = \gamma_F F_r \tag{8.1.2-1}$$

Where F_r——representative value of an action;

γ_F——partial factor of an action.

2 The design value of a material property, f_d, may be determined as follows:

$$f_d = \frac{f_k}{\gamma_M} \tag{8.1.2-2}$$

Where f_k——characteristic value of a material property;

γ_M——partial factor of a material property, whose value may be adopted according to the requirements specified in the relevant structure design standards.

3 The design value of a geometrical parameter, a_d, may be its characteristic value, a_k. The design value of a geometrical parameter may be determined as follows if the variability of the geometrical parameter has an obvious impact on the structural properties:

$$a_d = a_k \pm \Delta_a \tag{8.1.2-3}$$

Where Δ_a——additional quantity of a geometric parameter.

4 The design value of a structure resistance, R_d, may be determined as follows:

$$R_d = R(f_k/\gamma_M, a_d) \tag{8.1.2-4}$$

Note: The factor, γ_{Rd}, which reflects the uncertainty of the resistance model may be separated from the partial factor of a material property, γ_M, if it is required.

8.2 Ultimate Limit States

8.2.1 The following states shall be taken into account for the ultimate limit states design of the structure or structural member,:

1 Failure or excessive deformation of the structure or structural member (including the foundation), for which the strength of materials of the structure is critical;

2 The entire structure or parts of it, as a rigid body, loses its static equilibrium, for which the strength of the structural material or foundation is not critical;

3 Failure or excessive deformation of the foundation, for which the geotechnical strength is critical;

4 Fatigue failure of the structure or structural member, for which the material fatigue strength of the structure is critical.

8.2.2 The following requirements shall be met for the ultimate limit states design of the structure or structural member:

1 The following requirement shall be met for the ultimate limit states design of the structure or structural member (including the foundation) regarding failure or excessive deformation:

$$\gamma_0 S_d \leqslant R_d \tag{8.2.2-1}$$

Where γ_0——importance factor of the structure, whose value shall be adopted according to Appendix A;

S_d——design value of the effect of combination of actions (such as axial force, bending moment or vectors expressing with axial force and bending moment);

R_d——design value of resistance of the structure or structure member.

2 The following requirements shall be met for the ultimate limit states design of the entire structure or parts of it dealt with regarding losing static equilibrium as a rigid body:

$$\gamma_0 S_{d,dst} \leqslant S_{d,stb} \tag{8.2.2-2}$$

Where $S_{d,dst}$——design value of effect of an destabilishing action;

$S_{d,stb}$——design value of effect of an stabilishing action.

3 Partial factor method may be adopted for the ultimate limit states design of foundation regarding failure or excessive deformation. However, value of the partial factor is different from that in formula (8.2.2-1).

Note: The permissible stress method may also be used for the bearing capacity design of foundation regarding failure or excessive deformation.

4 The methods specified in Appendix F may be used for the ultimate limit states design of structure or structural member regarding fatigue failure.

8.2.3 The following requirements shall be met for combination of actions in the expressions of ultimate limit states design:

1 Combination of actions shall be the one for which the actions are likely to present simultaneously;

2 Each combination of actions shall include a leading variable action, or an accidental action, or a seismic action;

3 The favorable part and unfavorable part of the leading action shall be separated and taken as a single action respectively if the variation of the permanent action position in the structure is very sensitive to the static equilibrium or similar limit state;

4 The partial factor of the action which generates favorable effect shall be reduced if several kinds of effect generated by one action are not fully correlated ;

5 Different combinations of actions shall be adopted for different design situations.

8.2.4 The following basic combination of actions shall be adopted for the persistent design situation and transient design situation:

1 The design value of effect of a basic combination may be determined as follows:

$$S_d = S(\sum_{i\geqslant 1} \gamma_{G_i} G_{ik} + \gamma_P P + \gamma_{Q_1} \gamma_{L1} Q_{1k} + \sum_{j>1} \gamma_{Q_j} \psi_{cj} \gamma_{Lj} Q_{jk}) \tag{8.2.4-1}$$

Where $S(\cdot)$——function of effect of the combination of actions;

G_{ik}——characteristic value of the ith permanent action;

P——relevant representative value of a prestress action;

Q_{1k}——characteristic value of the first variable action (leading variable action);

Q_{jk}——characteristic value of the jth variable action;

γ_{G_i}——partial factor of the ith permanent action, which shall be adopted according to Appendix A;

γ_P——partial factor of a prestress action, which shall be adopted according to Appendix A;

γ_{Q_1}——the partial factor of the first variable action (leading variable action), which shall be adopted according to Appendix A;

γ_{Q_j}——partial factor of the jth variable action, which shall be adopted according to Appendix A;

γ_{L1}, γ_{Lj}——load adjustment factor of the first and the jth variable actions considering the design working life of the structure, which shall be adopted according to the relevant regulations. It shall be $\gamma_L=1.0$ for structures whose design working life is identical with the design reference period;

ψ_{cj}——combination value factor of the jth variable action, which shall be adopted according to those specified in the relevant codes.

Note: In the effect function of the combination of actions, $S(\cdot)$, symbols "Σ" and "+" both indicate the logical combination instead of the algebraic addition, that is, the joint influence of all the actions on the structure considered simultaneously.

2 The design value of the effect of basic combination may be calculated as follows if the relationship between an action and the effect of action are taken into account linearly:

$$S_d=\sum_{i\geqslant 1}\gamma_{G_i}S_{G_{ik}}+\gamma_P S_P+\gamma_{Q_1}\gamma_{L1}S_{Q_{1k}}+\sum_{j>1}\gamma_{Q_j}\psi_{cj}\gamma_{Lj}S_{Q_{jk}} \quad (8.2.4\text{-}2)$$

Where $S_{G_{ik}}$——characteristic value of the ith permanent action;

S_P——effect of the relevant representative value of a prestress action;

$S_{Q_{1k}}$——effect of the characteristic value of the first variable action (leading variable action);

$S_{Q_{jk}}$——effect of the characteristic value of the jth variable action.

Note: 1 The design value of effect of the combination of actions may also be given respectively as required for the persistent design situation and transient design situation;

2 The factor, γ_{Sd}, which reflects the uncertainty of the effect of action model may be separated from the partial factor of an action if it is required.

8.2.5 The accidental combination of actions shall be adopted for the accidental design situation:

1 The design value of effect of an accidental combination may be determined as follows:

$$S_d=S\left[\sum_{i\geqslant 1}G_{ik}+P+A_d+(\psi_{f1}\text{ or }\psi_{q1})Q_{1k}+\sum_{j>1}\psi_{qj}Q_{jk}\right] \quad (8.2.5\text{-}1)$$

Where A_d——design value of an accidental action;

ψ_{f1}——frequent value factor of the first variable action, which shall be adopted according to those specified in the relevant codes;

ψ_{q1}, ψ_{qj}——quasi-permanent value factor of the first and the jth variable actions, which shall be adopted according to those specified in the relevant codes.

2 The design value of effect for an accidental combination may be calculated as follows if the relationship between action and the effect of action are taken into account linearly:

$$S_d = \sum_{i\geqslant1} S_{G_{ik}} + S_P + S_{A_d} + (\psi_{f1} \text{ or } \psi_{q1}) S_{Q_{1k}} + \sum_{j>1} \psi_{qj} S_{Q_{jk}} \tag{8.2.5-2}$$

Where S_{A_d} ——effect of the design value of an accidental action.

8.2.6 The seismic combination of actions shall be adopted for seismic design situation.

1 The design value of effect of a seismic combination should be determined according to the seismic action (basic intensity) with return period of 475 years, and it shall meet the following requirements:

1) The design value of effect of a seismic combination should be determined as follows:

$$S_d = S(\sum_{i\geqslant1} G_{ik} + P + \gamma_I A_{Ek} + \sum_{j\geqslant1} \psi_{qj} Q_{jk}) \tag{8.2.6-1}$$

Where γ_I——importance factor of a seismic action, which shall be adopted according to those specified in the relevant codes for seismic design;

A_{Ek}——characteristic value of a seismic action determined according to the seismic action (basic intensity) with return period of 475 years.

2) The design value of effect of a seismic combination may be calculated as follows if the relationship between an action and the effect are taken into account linearly:

$$S_d = \sum_{i\geqslant1} S_{G_{ik}} + S_P + \gamma_I S_{A_{Ek}} + \sum_{j\geqslant1} \psi_{qj} S_{Q_{jk}} \tag{8.2.6-2}$$

Where $S_{A_{Ek}}$ ——effect of the characteristic value of a seismic action.

Note: The results shall be divided by the structural performance factor if the seismic effect is calculated using linear-elastic method in order to consider the influence of the structural ductility. The structural performance factor shall be adopted according to those specified in the relevant codes for seismic design.

2 The design value of effect of a seismic combination may also be determined according to a seismic action with return period larger or less than 475 years, and it shall be in accordance with those specified in the relevant codes for seismic design.

8.2.7 Values of the partial factor of a permanent action, γ_G, and partial factor of a prestress action, γ_P, in equation (8.2.4) shall not be larger than 1.0 if the permanent action effect or prestress action effect is favorable to the load carrying capacity of the structural member.

8.3 Serviceability Limit States

8.3.1 The following requirements shall be met for serviceability limit states design of the structure or structural member,:

$$S_d \leqslant C \tag{8.3.1}$$

Where S_d——design value of effect of a combination of actions (such as deformation and crack);

C——limit for deformation and crack in the design, which shall be adopted according to those specified in the relevant codes for structural design.

8.3.2 The characteristic combination, frequent combination or quasi-permanent combination of an action may be adopted for the serviceability limit states design according to different conditions.

1 Characteristic combination

1) The design value of effect of a characteristic combination may be determined as follows:

$$S_d = S(\sum_{i\geqslant1} G_{ik} + P + Q_{1k} + \sum_{j>1} \psi_{cj} Q_{jk}) \tag{8.3.2-1}$$

2) The design value of effect of a characteristic combination may be calculated as follows if the relationship between an action and the effect of action are taken into account linearly:

$$S_d = \sum_{i\geqslant 1} S_{G_{ik}} + S_p + S_{Q_{1k}} + \sum_{j>1} \psi_{cj} S_{Q_{jk}} \quad (8.3.2\text{-}2)$$

2 Frequent combination

1) The design value of effect of a frequent combination may be determined as follows:

$$S_d = S(\sum_{i\geqslant 1} G_{ik} + P + \psi_{f1} Q_{1k} + \sum_{j>1} \psi_{qj} Q_{jk}) \quad (8.3.2\text{-}3)$$

2) The design value of effect of a frequent combination may be calculated as follows if the relationship between an action and the effect are taken into account linearly:

$$S_d = \sum_{i\geqslant 1} S_{G_{ik}} + S_p + \psi_{f1} S_{Q_{1k}} + \sum_{j>1} \psi_{qj} S_{Q_{jk}} \quad (8.3.2\text{-}4)$$

3 Quasi-permanent combination

1) The design value of effect of a quasi-permanent combination may be determined as follows:

$$S_d = S(\sum_{i\geqslant 1} G_{ik} + P + \sum_{j\geqslant 1} \psi_{qj} Q_{jk}) \quad (8.3.2\text{-}5)$$

2) The design value of effect of a quasi-permanent combination may be calculated as follows if the relationship between an action and the effect are taken into account linearly:

$$S_d = \sum_{i\geqslant 1} S_{G_{ik}} + S_p + \sum_{j\geqslant 1} \psi_{qj} S_{Q_{jk}} \quad (8.3.2\text{-}6)$$

Note: The characteristic combination should be used for irreversible serviceability limit states; the frequent combination should be used for reversible serviceability limit states; while the quasi-permanent combination should be used for serviceability limit states when the long-term effect is the decisive factor.

8.3.3 The partial factor of a material property, γ_M, shall be 1.0 in the serviceability limit states verification unless there is specific regulation in the codes for structural design.

Appendix A Special Provisions for Various Engineering Structures

A. 1 Special Provisions for Building Structures

A. 1. 1 The safety of the building structure shall be classified according to the consequence of failure of the structure as shown in Table A. 1. 1.

Table A. 1. 1 Safety Classes of Building Structures

Safety Classes	Consequence of failure	Examples
Class Ⅰ	Very serious: having a large impact on the human life, economy, society or environment	Large-scale public buildings
Class Ⅱ	Serious: having an impact on the human life, economy, society or environment	Ordinary residential buildings and office buildings
Class Ⅲ	Less serious: having a little impact on the human life, economy, society or environment	Small or temporary storage buildings

Note: The safety class for the buildings of category A and class B in seismic design should be Ⅰ; the safety class for the buildings of category C should be Ⅱ; and the safety class for the buildings of category D should be Ⅲ.

A. 1. 2 The design reference period of building structures is 50 years.

A. 1. 3 The design working life of the building structures shall be adopted according to Table A. 1. 3.

Table A. 1. 3 Design Working Life of Building Structures

Classifications	Design working life (years)	Examples
1	5	Temporary building structures
2	25	Easy replaceable structural members
3	50	Ordinary buildings and special structures
4	100	Landmark buildings and very important building structures

A. 1. 4 The reliability index of ultimate limit states design of building structural members in the persistent design situation shall not be less than those specified in Table A. 1. 4.

Table A. 1. 4 Reliability Index, β of building Structural Members

Types of failure	Safety classes		
	Class Ⅰ	Class Ⅱ	Class Ⅲ
Ductile	3.7	3.2	2.7
Brittle	4.2	3.7	3.2

A. 1. 5 The reliability index of serviceability limit states design of building structureal members in the persistent design situation should be in a range of 0～1.5 based on its reversibility.

A. 1. 6 The following requirements shall be complied with for the persistent design situation and transient design situation in the ultimate limit states design:

1 The design value of effect of a combination of actions shall be the unfavorable value

obtained from equation (8. 2. 4-1) and the following equation:

$$S_d = S(\sum_{i\geq 1} \gamma_{G_i} G_{ik} + \gamma_P P + \gamma_L \sum_{j\geq 1} \gamma_{Q_j} \psi_{cj} Q_{jk}) \tag{A. 1. 6-1}$$

2 The design value of effect of a combination of actions shall be the unfavorable value obtained from equation (8. 2. 4-2) and the following equation if the relationship between an action and an effect are taken into account linearly:

$$S_d = \sum_{i\geq 1} \gamma_{G_i} S_{G_{ik}} + \gamma_P S_P + \gamma_L \sum_{j\geq 1} \gamma_{Q_j} \psi_{cj} S_{Q_{jk}} \tag{A. 1. 6-2}$$

A. 1. 7 The importance factor, γ_0, of a building structure shall not be less than those specified in Table A. 1. 7.

Table A. 1. 7 Importance factor, γ_0, of a Building Structure

importance factor	The persistent design situation and transient design situation			The accidental design situation and seismic design situation
	Safety classes			
	Class Ⅰ	Class Ⅱ	Class Ⅲ	
γ_0	1. 1	1. 0	0. 9	1. 0

A. 1. 8 The partial factor of an action of building structure shall be in accordance with those specified in Table A. 1. 8.

Table A. 1. 8 Partial Factor of an Action of a Building Structure

Applicable condition / Partial factor of an action	The effect of action is unfavorable		The effect of action is favorable
	For equation (8. 2. 4-1) and (8. 2. 4-2)	For equation(A. 1. 6-1) and (A. 1. 6-2)	
γ_G	1. 2	1. 35	≤1. 0
γ_P	1. 2		1. 0
γ_Q	1. 4		0

A. 1. 9 The load adjustment factor considering the design working life of the building structures shall be in accordance with those specified in Table A. 1. 9.

Table A. 1. 9 Load Adjustment factor, γ_L, Considering the Design Working Life of the Building Structures

Design working life (years)	γ_L
5	0. 9
50	1. 0
100	1. 1

Note: γ_L shall be adopted according to those specified in the design codes for various material structures for structural members whose design working life is 25 years.

A. 2 Special Provisions for Railway Bridge and Culvert Structures

A. 2. 1 The safety class of railway bridge and culvert structures is class I.

A. 2. 2 The design reference period of railway bridge and culvert structures is 100 years.

A. 2. 3 The design working life of railway bridge and culvert structures is 100 years.

A. 2. 4 The basic combination and accidental combination of actions shall be adopted for the ultimate limit states design of railway bridge and culvert structures.

1 Basic combination

1) The design value of effect of a basic combination shall be determined as follows:

$$S_d = \gamma_{Sd} S(\sum_{i \geqslant 1} \gamma_{G_i} G_{ik} + \gamma_{Q_1} Q_{1k} + \sum_{j>1} \gamma_{Q_j} Q_{jk}) \quad \text{(A. 2. 4-1)}$$

Where γ_{Sd}——uncertainty factor of an action model, which is generally 1.0;

$S(\cdot)$——effect function of a combination of actions, in which symbols "$\sum$" and "+" both express logical combination;

G_{ik}——characteristic value of the ith permanent action;

Q_{1k}, Q_{jk}——characteristic value of the first(leading) and the jth variable action;

γ_{G_i}——partial factor of the ith permanent action;

γ_{Q_1}、γ_{Q_j}——partial factor of a combination of the first(leading) and the jth variable actions.

2) The design value of effect of a basic combination may be calculated as follows if the relationship between an action and the effect of action are taken into account linearly:

$$S_d = \gamma_{Sd} (\sum_{i \geqslant 1} \gamma_{G_i} S_{G_{ik}} + \gamma_{Q_1} S_{Q_{1k}} + \sum_{j>1} \gamma_{Q_j} S_{Q_{jk}}) \quad \text{(A. 2. 4-2)}$$

Where $S_{G_{ik}}$——characteristic value of the ith permanent action;

$S_{Q_{1k}}$、$S_{Q_{jk}}$——effect of the characteristic value of the first (leading) and the jth variable action.

2 Accidental combination

1) The design value of effect of an accidental combination may be determined as follows:

$$S_d = S(\sum_{i \geqslant 1} G_{ik} + A_d + \sum_{j \geqslant 1} \gamma_{Q_j} Q_{jk}) \quad \text{(A. 2. 4-3)}$$

Where A_d——the design value of an accidental action.

2) The design value of effect of an accidental combination may be calculated as follow if the relationship between an action and the effect of action are taken into account linearly

$$S_d = \sum_{i \geqslant 1} S_{G_{ik}} + S_{A_d} + \sum_{j \geqslant 1} \gamma_{Q_j} S_{Q_{jk}} \quad \text{(A. 2. 4-4)}$$

Where S_{A_d}——effect of the design value of an accidental action.

A. 2. 5 The characteristic combination of actions shall be adopted for the serviceability limit states design of railway bridge and culvert structures.

1 The design value of effect of characteristic combination shall be determined as follows:

$$S_d = \gamma_{Sd} S(\sum_{i \geqslant 1} G_{ik} + Q_{1k} + \sum_{j>1} \gamma_{Q_j} Q_{jk}) \quad \text{(A. 2. 5-1)}$$

Where γ_{Q_j}——partial factor of the jth variable action in the serviceability limit states design.

2 The design value of effect of characteristic combination shall be calculated as follows if the relationship between an action and the effect of action are taken into account linearly:

$$S_d = \gamma_{Sd} (\sum_{i \geqslant 1} S_{G_{ik}} + S_{Q_{1k}} + \sum_{j>1} \gamma_{Q_j} S_{Q_{jk}}) \quad \text{(A. 2. 5-2)}$$

A. 2. 6 The following limits shall be established according to the line grades and bridge types for the serviceability limit states design of railway bridge and culvert structures:

1 Limits of vertical deflection, beam end angle and vertical natural vibration frequency of the bridge span structures under the static live loads;

2 Limits of transverse wide to span ratioand horizontal displacement of the bridge span structures and limit of transverse vibration frequency of the whole bridge;

3 Vehicle-bridge coupling dynamic response analysis shall be conducted for the bridge structures in lines on which the travelling speed of trains is less than 200km/h. And the train traveling shall meet the safety and comfort requirements;

4 Limit of crack width of reinforced concrete and partial prestressing members which are allowed to crack occur under different corrosive environment;

5 Influence of the rigidity fatigue reduction factor on the rigidity of the member shall be taken into account in the deformation calculation of bending concrete members.

A. 2. 7 Fatigue bearing capacity shall be verified according to the following requirements for the welded or un-welded tension or tension-compression steel structural members as well as the concrete bending members withstanding the repeated stress generated by the trains in the railway bridge and culvert structures:

1 The typical fatigue train and fatigue action (stress) spectrum, as well as the proof load effect spectrum established according to survey and statistical analysis for lines with different traffic volume grades may be used for the fatigue loads of railway bridge and culvert structures.

2 The equivalent uniform amplitude repeated stress method should be used for the fatigue ultimate limit states verification of the railway bridge and culvert structures.

A. 3 Special Provisions for Highway Bridge and Culvert Structures

A. 3. 1 The safety of highway bridge and culvert structures shall be classified in accordance with the specification in Table A. 3. 1.

Table A. 3. 1 Safety Classes of Highway Bridge and Culvert Structures

Safety classes	Types	Examples
Class Ⅰ	Important	Grand bridges, large bridges, intermediate bridge and important small bridge
Class Ⅱ	Ordinary	Small bridge, important culverts and important retaining walls
Class Ⅲ	Secondary	Culverts, retaining walls and Collision guard fences

A. 3. 2 The design reference period of railway bridge and culvert structures is 100 years.

A. 3. 3 The design working life of the highway bridge and culvert structures shall be in accordance with the specification in Table A. 3. 3.

Table A. 3. 3 Design Working Life of Highway Bridge and Culvert Structure

Types	Design working life (years)	Examples
1	30	Small bridge and culverts
2	50	Intermediate bridge and important small bridge
3	100	Grand bridge, large bridge and important intermediate bridge

Note: The design working life of structures with special requirements may be adjusted through technical and economic verification based on Table A. 3. 3.

A. 3. 4 The basic combination of actions shall be adopted for the persistent design situation and transient design situation in the ultimate limit states design of highway bridge and culvert structures; while accidental combination of actions shall be adopted for the accidental design situation.

1 Basic combination

1) The design value of effect of basic combination, S_d, may be determined as follows:

$$S_d = S(\sum_{i\geq 1} \gamma_{G_i} G_{ik} + \gamma_{Q_1} \gamma_L Q_{1k} + \psi_c \gamma_L \sum_{j>1} \gamma_{Q_j} Q_{jk}) \quad (A.3.4\text{-}1)$$

Where $S(\cdot)$——effect function of a combination of actions, in which symbols "$\sum$" and "+" both express logical combination;

G_{ik}——characteristic value of the ith permanent action;

Q_{1k}——characteristic value of the first variable action (leading variable action);

Q_{jk}——characteristic value of the jth variable action;

γ_{G_i}——partial factor of the ith permanent action, which shall be adopted in accordance with Table A. 3. 7;

γ_{Q_1}——partial factor of the first variable action (leading variable action), which shall be adopted according to those specified in the relevant codes for highway bridge and culvert structures;

γ_{Q_j}——partial factor of the jth variable action, which shall be adopted according to the specification in the relevant codes for highway bridge and culvert structures;

γ_L——load adjustment factor considering the design working life of the structure, which shall be in accordance with the specification in the relevant codes for highway bridge and culvert structures;

ψ_c——combination value factor of a variable action, which shall be adopted according to the specification in the relevant codes for highway bridge and culvert structures.

2) The design value of effect of basic combination S_d may be calculated as follows if the relationship between an action and the effect of action are taken into account linearly:

$$S_d = \sum_{i\geq 1} \gamma_{G_i} S_{G_{ik}} + \gamma_{Q_1} \gamma_L S_{Q_{1k}} + \psi_c \gamma_L \sum_{j>1} \gamma_{Q_j} S_{Q_{jk}} \quad (A.3.4\text{-}2)$$

Where $S_{G_{ik}}$——characteristic value of the ith permanent action;

$S_{Q_{1k}}$——effect of the characteristic value of the first variable action (leading variable action);

$S_{Q_{jk}}$——effect of the characteristic value of the jth variable action.

2 Accidental combination

1) The design value of effect of an accidental combination, S_d, may be determined as follows:

$$S_d = S(\sum_{i\geq 1} G_{ik} + A_d + (\psi_{f1} \text{ or } \psi_{q1}) Q_{1k} + \sum_{j>1} \psi_{qj} Q_{jk}) \quad (A.3.4\text{-}3)$$

Where A_d——design value of an accidental action;

ψ_{f1}——frequent value factor of the first (leading) variable action, which shall be adopted according to the specification in the relevant codes for highway bridge and culvert structures codes;

ψ_{q1}, ψ_{qj}——quasi-permanent value factor of the first (leading) and the jth variable actions, which shall be adopted according to the specification in the relevant codes for highway bridge and culvert structures.

2) The design value of effect of an accidental combination may be calculated as follows if the relationship between an action and the effect of action are taken into account linearly:

$$S_d = \sum_{i\geq 1} S_{G_{ik}} + S_{A_d} + (\psi_{f1} \text{ or } \psi_{q1}) S_{Q_{1k}} + \sum_{j>1} \psi_{qj} S_{Q_{jk}} \qquad \text{(A. 3. 4-4)}$$

Where S_{A_d} ——effect of the design value of an accidental action.

A. 3. 5 The characteristic combination, frequent combination or quasi-permanent combination of actions for the serviceability limit states design of highway bridge and culvert structures shall be adopted according to different conditions.

1 Characteristic combination

1) The design value of effect of a characteristic combination, S_d, may be determined as follows:

$$S_d = S(\sum_{i\geq 1} G_{ik} + Q_{1k} + \psi_c \sum_{j>1} Q_{jk}) \qquad \text{(A. 3. 5-1)}$$

2) The design value of effect of a characteristic combination, S_d, may be calculated as follows if the relationship between an action and the effect of action are taken into account linearly:

$$S_d = \sum_{i\geq 1} S_{G_{ik}} + S_{Q_{1k}} + \psi_c \sum_{j>1} S_{Q_{jk}} \qquad \text{(A. 3. 5-2)}$$

2 Frequent combination

1) The design value of effect of a frequent combination, S_d, may be determined as follows:

$$S_d = S(\sum_{i\geq 1} G_{ik} + \psi_{f1} Q_{1k} + \sum_{j>1} \psi_{qj} Q_{jk}) \qquad \text{(A. 3. 5-3)}$$

2) The design value of effect of a frequent combination, S_d, shall be calculated as follows if the relationship between an action and the effect of action are taken into account linearly:

$$S_d = \sum_{i\geq 1} S_{G_{ik}} + \psi_{f1} S_{Q_{1k}} + \sum_{j>1} \psi_{qj} S_{Q_{jk}} \qquad \text{(A. 3. 5-4)}$$

3 Quasi-permanent combination

1) The design value of effect of a quasi-permanent combination S_d may be determined as follows:

$$S_d = S(\sum_{i\geq 1} G_{ik} + \sum_{j\geq 1} \psi_{qj} Q_{jk}) \qquad \text{(A. 3. 5-5)}$$

2) The design value of effect of a quasi-permanent combination, S_d, shall be calculated as follows if the relationship between an action and the effect of action are taken into account linearly:

$$S_d = \sum_{i\geq 1} S_{G_{ik}} + \sum_{j\geq 1} \psi_{qj} S_{Q_{jk}} \qquad \text{(A. 3. 5-6)}$$

A. 3. 6 The importance factor of a highway bridge and culvert structure shall not be less than the value in Table A. 3. 6.

Table A. 3. 6 Importance Factor, γ_0, of a Highway Bridge and Culvert Structure

Safety classes	Class Ⅰ	Class Ⅱ	Class Ⅲ
Importance Factor γ_0	1. 1	1. 0	0. 9

A. 3. 7 The partial factor of a permanent action of a highway bridge and culvert structure shall be adopted according to Table A. 3. 7.

Table A. 3. 7 Partial Factor, γ_G, of a Permanent Action of Highway Bridge and Culvert Structure

Serial number	Actions		The effect of action is unfavorable	The effect of action is favorable
1	Gravity of concrete and masonry structures (including additional gravity of the structure)		1. 2	1. 0
	Gravity of steel structures (including additional gravity of the structure)		1. 1～1. 2	
2	Prestress		1. 2	
3	Soil gravity		1. 2	
4	Concrete shrinkage and creep action		1. 0	
5	Lateral earth pressure		1. 4	
6	Buoyancy of water		1. 0	
7	Action caused by foundation displacement	Concrete and masonry structures	0. 5	0. 5
		Steel structures	1. 0	1. 0

A. 4 Special Provisions for Port and Harbor Structures

A. 4. 1 The safety classes of port and harbor structures shall be classified in accordance with Table A. 4. 1.

Table A. 4. 1 Safety Classes of Port and harbor Structuresp

Safety classes	Failure consequence	Applicability
Class Ⅰ	Very serious	Structures with special safety requirements
Class Ⅱ	Serious	Ordinary port and harbor structures
Class Ⅲ	Less serious	Temporary port and harbor structures

A. 4. 2 The design reference period of port and harbor structures is 50 years.

A. 4. 3 The design working life of port and harbor structures shall be in accordance with Table A. 4. 3.

Table A. 4. 3 Classifications of Design Working Life of Port and Harbor Structures

Types	Design working life (years)	Examples
1	5～10	Temporary port and harbor structures
2	50	Permanent and harbor port structures

A. 4. 4 The reliability index of port and harbor structures for the persistent design situation in the ultimate limit states design should not be less than the value shown in Table A. 4. 4.

Table A. 4. 4 Reliability Index of Port and harbor Structures

Structures	Safety classes		
	Class Ⅰ	Class Ⅱ	Class Ⅲ
Ordinary port and harbor structures	4. 0	3. 5	3. 0

Note: Exclusive of earth slopes and foundation stability/breakwater structures.

A. 4. 5 The permanent, transient, accidental and seismic combination of actions shall be adopted for the ultimate limit state design according to different design situations.

1 Permanent combination

1) The design value of effect should be determined as follows for the permanent combinations of actions of the port and harbor structures:

$$S_d = S(\sum_{i\geqslant 1} \gamma_{G_i} G_{ik} + \gamma_P P + \gamma_{Q_1} Q_{1k} + \sum_{j>1} \gamma_{Q_j} \psi_{cj} Q_{jk}) \quad \text{(A. 4. 5-1)}$$

Where $S(\cdot)$——effect function of a combination of actions, in which symbols "$\sum$" and "+" both express logical combination;

G_{ik}——characteristic value of the ith permanent action;

P——representative value of a prestress;

Q_{1k}, Q_{jk}——characteristic value of the first (leading) and the jth variable action;

γ_{G_i}——partial factor of the ith permanent action, whose value shall be in accordance with those specified in Table A. 4. 12;

γ_P——partial factor of a prestress;

γ_{Q_1}, γ_{Q_j}——partial factor of the first (leading) and the jth variable actions, whose value shall be in accordance with those specified in Table A. 4. 12;

ψ_{cj}——combination value factor of a variable action, which may be 0. 7; for bounded action presenting with its bounded values frequently, it may be 1. 0.

2) The design value of effect of a permanent combination of actions may be calculated as follows if the relationship between an action and the effect of action are taken into account linearly:

$$S_d = \sum_{i\geqslant 1} \gamma_{G_i} S_{G_{ik}} + \gamma_P S_P + \gamma_{Q_1} S_{Q_{1k}} + \sum_{j>1} \gamma_{Q_j} \psi_{cj} S_{Q_{jk}} \quad \text{(A. 4. 5-2)}$$

3) The design value of effect may also be determined as follows for the permanent combination of actions in some cases:

$$S_d = \gamma_F S(\sum_{i\geqslant 1} G_{ik} + \sum_{j\geqslant 1} Q_{jk}) \quad \text{(A. 4. 5-3)}$$

Where γ_F——lumped partial factor of actions, which shall be specified in the relevant design codes.

2 Transient combination

1) The design value of effect should be determined as follows for the transient combination of actions of port and harbor structures:

$$S_d = S(\sum_{i\geqslant 1} \gamma_{G_i} G_{ik} + \gamma_P P + \sum_{j\geqslant 1} \gamma_{Q_j} Q_{jk}) \quad \text{(A. 4. 5-4)}$$

2) The design value of effect of a transient combination may be calculated as follows if the relationship between an action and the effect of action are taken into account linearly:

$$S_d = \sum_{i\geqslant 1} \gamma_{G_i} S_{G_{ik}} + \gamma_P S_P + \sum_{j\geqslant 1} \gamma_{Q_j} S_{Q_{jk}} \quad \text{(A. 4. 5-5)}$$

Where γ_{Q_j}——partial factor of the jth variable action, which may be 0. 1 decreased from the values specified in Table A. 4. 12.

3) The design value of effect may also be determined according to the equation (A. 4. 5-3) for the transient combination of actions in some cases.

3 Accidental combination

The following requirements shall be met for the accidental combination:

1) The partial factor of an accidental action shall be 1. 0;

2) The characteristic value shall be adopted for variable action appearing with the accidental action simultaneously.

4 Seismic combination

The following requirements shall be met for the seismic combination:

1) Partial factor of the representative value of a seismic action shall be 1. 0;

2) The specific design expressions and factors shall be in accordance with the specification in the current relevant national standards.

A. 4. 6 The characteristic, frequent and quasi-permanent combination of actions may be respectively adopted for the serviceability limit states verification of the persistent design situation according to different design requirements, and the requirement of equation (8. 3. 1) should be satisfied.

1 Characteristic combination

1) The design value of effect of a characteristic combination may be determined as follows:

$$S_d = S(\sum_{i\geqslant 1} G_{ik} + P + Q_{1k} + \sum_{j>1} \psi_{cj} Q_{jk}) \quad (A. 4. 6\text{-}1)$$

2) The design value of effect of a characteristic combination may be calculated as follows if the relationship between an action and the effect of action are taken into account linearly:

$$S_d = \sum_{i\geqslant 1} S_{G_{ik}} + S_P + S_{Q_{1k}} + \sum_{j>1} \psi_{cj} S_{Q_{jk}} \quad (A. 4. 6\text{-}2)$$

2 Frequent combination

1) The design value of effect of a frequent combinationcan be determined as follows:

$$S_d = S(\sum_{i\geqslant 1} G_{ik} + P + \psi_f Q_{1k} + \sum_{j>1} \psi_{qj} Q_{jk}) \quad (A. 4. 6\text{-}3)$$

2) The design value of effect of a frequent combination may be calculated as follows if the relationship between an action and the effect of action are taken into account linearly:

$$S_d = \sum_{i\geqslant 1} S_{G_{ik}} + S_P + \psi_f S_{Q_{1k}} + \sum_{j>1} \psi_{qj} S_{Q_{jk}} \quad (A. 4. 6\text{-}4)$$

3 Quasi-permanent combination

1) The design value of effect of a quasi-permanent combination may be determined as follows:

$$S_d = S(\sum_{i\geqslant 1} G_{ik} + P + \sum_{j\geqslant 1} \psi_{qj} Q_{jk}) \quad (A. 4. 6\text{-}5)$$

2) The design value of effect of a quasi-permanent combination may be calculated as follows if the relationship between an action and the effect of action are taken into account linearly:

$$S_d = \sum_{i\geqslant 1} S_{G_{ik}} + S_P + \sum_{j\geqslant 1} \psi_{qj} S_{Q_{jk}} \quad (A. 4. 6\text{-}6)$$

Where ψ_{cj}, ψ_{f}, ψ_{qj}——combination, frequent and quasi-permanent value factor of a variable action respectively.

A. 4. 7 The design water level of the port and harbor structure shall be determined according to the following requirements for the combination of actions in the ultimate limit state design:

1 Permanent combination: calculate respectively for the design high/low water level, extreme high/low water level, and a certain unfavorable water level between the design high water level, and in combination with the underground water level;

2 Transient combination: calculate respectively for the design high/low water level and a certain unfavorable water level between the design high water level, and in combination with the underground water level.

A. 4. 8 The design water level of the river port and harbor shall be determined according to the following requirements for the combination of actions in the ultimate limit state design:

1 Permanent combination: calculate respectively for the design high/low water level and a certain unfavorable water level, and in combination with the underground water level;

2 Transient combination: calculate respectively for the design high water level and design low water level; a certain unfavorable water level may be taken for construction period.

A. 4. 9 The design water level for the seismic combination shall be in accordance with the specification in the current relevant standards of the nation in the ultimate limit state design.

A. 4. 10 The extreme water level may not be taken into account for the combination of actions adopted in the serviceability limit states design.

A. 4. 11 The importance factor of port and harbor structures shall be in accordance with Table A. 4. 11.

Table A. 4. 11 Importance Factor of Port and Harbor Structures

Safety classes	Class Ⅰ	Class Ⅱ	Class Ⅲ
Importance Factor γ_0	1. 1	1. 0	0. 9

Note: 1 γ_0 may be appropriately increased for port and harbor structure with safety class Ⅰ and special requirements for safety required;

2 γ_0 may be appropriately increased if the natural conditions are complex and the structure is difficult for maintaining.

A. 4. 12 The partial factor of an action for the permanent combination in the ultimate limit state design shall be in accordance with Table A. 4. 12.

Table A. 4. 12 Partial Factor of an Action

Actions	Partial factors	Actions	Partial factors
Permanent load (excluding the earth pressure and hydrostatic pressure)	1. 2	Railway load	1. 4
Hardware steel load	1. 5	Vehicle load	1. 4
Bulk-load load		Cable car load	
Hoisting machinery load		Ship mooring force	
Ship impact force		Ship breasting force	
Flow force		Transport machinery load	
Ice load		Wind load	
Wave force (for member calculation)		Crowd load	

TableA. 4. 12(Continued)

Actions	Partial factors	Actions	Partial factors
General cargo and container load	1. 4	Earth pressure	1. 35
Fluid pipeline (including the thrust) load		Residual water pressure	1. 05

Note: 1 The partial factor of a permanent action γ_G shall not be larger than 1. 0 if the permanent action effect is favorable to the load carrying capacity of the structure;

2 Each sub-action shall multiply by the partial factor of the unfavorable action for action of the same source if the resultant action is unfavorable to the load carrying capacity of the structure;

3 If the permanent action is the leading one, its partial factor shall not be less than 1. 3;

4 The partial factor of the accompanying variable action shall be taken the same with the leading variable action if two variable actions are completely correlated and one of them is the leading;

5 The partial factor of a variable action of the permanent combination in the ultimate limit states shall be 0. 1 decreased for the sea port and harbor structure in the extreme high water level and extreme low water level;

6 The partial factor of the wave force in the calculation against slide and overturning shall be in accordance with the specification in the relevant structural codes.

Appendix B Quality Control

B.1 Requirements for Quality Control

B.1.1 The quality of materials and members may be expressed by one or more qualitative characteristics. Definite requirements for the qualitative characteristics of the materials and members such as mechanical properties and geometrical parameter shall be specified in the structural design and construction codes.

The acceptable quality level of materials and members shall be determined according to the reliability index of structural member specified in the related codes of various engineering structures.

B.1.2 The materials should be graded according to the statistical data and different quality levels. The division of grade should not be over narrowed. Different characteristic values of material property shall be adopted for different grade materials in the design of structure.

B.1.3 Essential quality control measure shall be taken for the engineering structureto fulfill the specified structural reliability. The quality control requirements of engineering structure shall be specified in related standards and the following contents shall be covered:

1 Quality control for reconnaissance and design;

2 Quality control for materials and products;

3 Quality control for construction;

4 Quality control for operation and maintenance.

B.1.4 The following requirements shall be met for the quality control of reconnaissance and design:

1 The survey information shall comply with the engineering requirements for the accuracy of the data;

2 The design scheme, fundamental assumption and computational model are reasonable and valid;

3 The drawings and other design documents comply with the relevant regulations.

B.1.5 Quality self-checking shall be exercised in each procedure and inspection shall be made for interchange between two procedures to control the construction quality. For the quality in operation and of the intermediate product, sampling test shall be carried out and analyzed by statistical method; and systematic inspection shall be carried out for the important part of structure.

B.1.6 The following two aspects of quality control for materials and members shall be covered:

1 Production control: routine inspection shall be made for material properties and member performance according to the relevant control standards, correct the deviations and keep the stability of quality of products in the producing process;

2 Compliance control (acceptance): acceptance inspection shall be carried out for materials and members according to relevant quality control standard before putting into service for ensuring the quality meets the requirement.

B. 1. 7 The compliance control may be carried out by sampling inspection method.

The specific quality acceptance standard for the materials and members shall be formulated according to their characteristics, including the acceptance inspection lot, sampling method and quantity, acceptance function, and acceptance criteria.

The quality acceptance standard should be formulated based on the statistic theory.

B. 1. 8 The risk of user must be controlled in formulation of the quality acceptance standard for the materials and member with poor manufacturing continuity or large difference in the statistical parameters of the qualitative characteristics among the batches. The limiting quality level used for calculation of the risk of user may be determined according to the related requirements in the design code of structures combined with engineering experience.

The quality acceptance standard may be formulated according to the risk of the producer of the materials and members only in the case that the production keeps continuation and the product quality is stable.

B. 1. 9 Reinspection or redetermination for the same quality grade shall be made according to related quality acceptance standard or be treated by other measures in the case that one batch of materials or members are unacceptable in the sampling inspection.

B. 2 Design Inspection and Construction Supervision

B. 2. 1 The engineering structure shall be subjected to the design inspection and construction supervision, and the associated requirements shall comply with the relevant regulations.

Note: At least two sets of computer software that the computational model comply with engineering practice shall be adopted for important or complicated engineering , and the computational solution shall be analyzed and compared carefully and then used for engineering design until it is ascertained to be reasonable and correct.

Appendix C Action Examples and Principle for Determination of Representative Value of a Variable Action

C.1 Action Examples

C.1.1 The permanent actions may include:

1 Dead load;

2 Earth pressure;

3 Water pressure at constant water level;

4 Prestress;

5 Foundation deformation;

6 Concrete shrinkage;

7 Welding deformation of steel products;

8 Construction factors resulting in imposed deformation or constrained deformation of structure.

C.1.2 The variable actions may include:

1 The load of staff and articles in operation;

2 Certain deadweight of structure in construction;

3 Installation load;

4 Vehicular load;

5 Crane load;

6 Wind load;

7 Snow load;

8 Ice load;

9 Seismic action;

10 Impact;

11 Water pressure at different water levels;

12 Uplift pressure;

13 Wave force;

14 Temperature variation.

C.1.3 The accidental actions may include:

1 Impact;

2 Explosion;

3 Seismic action;

4 Tornado;

5 Fire hazard;

6 Super-severe corrosion;

7 Flood action.

Note: The seismic action and impact may be considered either as the variable action under specific conditions or as the accidental action.

C. 2 Principle for Determination of the Representative Value of a Variable Action

C. 2. 1 The characteristic value of a variable action may be determined according to the following principle:

1 The probability distribution function, $F_T(x)$, of the maximum value of variable action in design reference period T may be determined according to the following equation if the stationary binomial process model is used:

$$F_T(x) = [F(x)]^m \tag{C. 2. 1-1}$$

Where $F(x)$ ——arbitrary-time-point probability distribution function of stochastic process of variable action;

m——average occurrence number of variable action in design reference period T.

If the arbitrary-time-point probability distribution is the extreme value I-type distribution (for example, the annual maximum wind pressure) as follows:

$$F(x) = \exp\left[-\exp\left(-\frac{x-u}{\alpha}\right)\right] \tag{C. 2. 1-2}$$

the probability distribution function of its maximum value is:

$$F_T(x) = \exp\left\{-\exp\left[-\frac{x-(u+\alpha\ln m)}{\alpha}\right]\right\} \tag{C. 2. 1-3}$$

2 The characteristic value, Q_k, of a variable action may be determined by the statistical characteristics of the maximum value probability distribution of variable action in the design reference period, T. Typically, the mean, the median, the mode, and the fractile, as follows, of specified probability, p, may be used:

$$F_T(Q_k) = p \tag{C. 2. 1-4}$$

In the situation of fractile, the probability of the maximum value exceeding characteristic value, Q_k, in the design reference period is $1-p$.

3 It is more convenient to adopt the return period, T_R, to express the characteristic value, Q_k, of a variable action in many cases especially for the nature action. The return period refers to the average interval in which two immediate appeared actions exceed Q_k. The relationship between Q_k and T_R is as follows:

$$F(Q_k) = 1 - 1/T_R \tag{C. 2. 1-5}$$

The following approximate relationship is existed among return period, T_R, probability, p, and the design reference period, T:

$$T_R \approx \frac{1}{\ln(1/p)} T \tag{C. 2. 1-6}$$

C. 2. 2 The frequent value of a variable action may be determined according to the following principle:

1 The frequent value may be determined according to the specified ratio of the total duration when the characteristic value of action is exceeded to the design reference period.

In the analysis of the stochastic process of a variable action, the ratio $\eta_x = T_x/T$, that is, the total duration $T_x = \sum_{i\geqslant 1} t_i$ when the value of an action exceeding a certain level Q_x divided by the design reference period, T, is used to represent the transiency of frequent value of an action (Figure C. 2. 2-1a). Figure C. 2. 2-1b presents the probability distribution function, $F_{Q^*}(x)$, of the value Q^* of variable action Q at arbitrary-time-point in the nonzero time domain. The probability, p^*, exceeding level Q_x may be determined according to the following equation:

$$p^* = 1 - F_{Q^*}(Q_x) \tag{C. 2. 2-1}$$

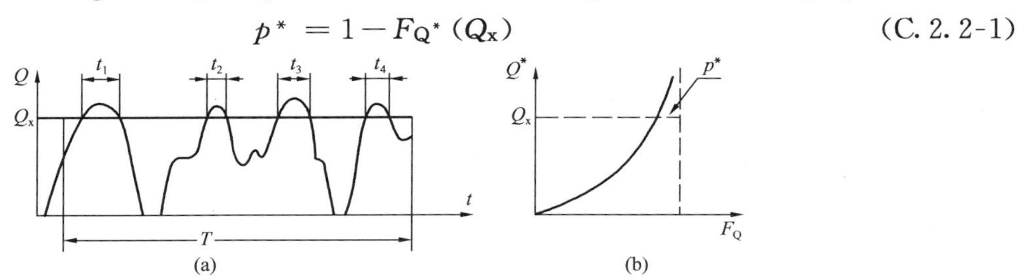

Figure C. 2. 2-1 Frequent Value of a Variable Action Defined According to the Ratio of the Total Duration when the Value of an Action Exceeding Certain Level Q_x to the Design Reference Period T

The following relationship is valid for the ergodic stochastic process:

$$\eta_x = p^* q \tag{C. 2. 2-2}$$

Where q——nonzero probability of action Q.

The action level Q_x may be determined as follows when η_x is specified:

$$Q_x = F_{Q^*}^{-1}\left(1 - \frac{\eta_x}{q}\right) \tag{C. 2. 2-3}$$

The frequent value of action for the serviceability limit states related to time may be determined according to this method. If a certain limit states is allowed being exceeded in a short duration or in a short time as a whole, smaller η_x value (not less than 0. 1) may be adopted, and the frequent value $\psi_f Q_k$ of the action shall be calculated according to Equation (C. 2. 2-3).

2 The frequent value may be determined according to the total number of value of action being exceeded or the average exceeding number in unit time interval (crossing rate). In the analysis of the stochastic process of variable action, the number n_x that the value of an action exceeds certain level Q_x or the average exceeding number $v_x = n_x/T$ (crossing rate) in unit time interval are used to represent the density that the frequent value appears (Figure C. 2. 2-2).

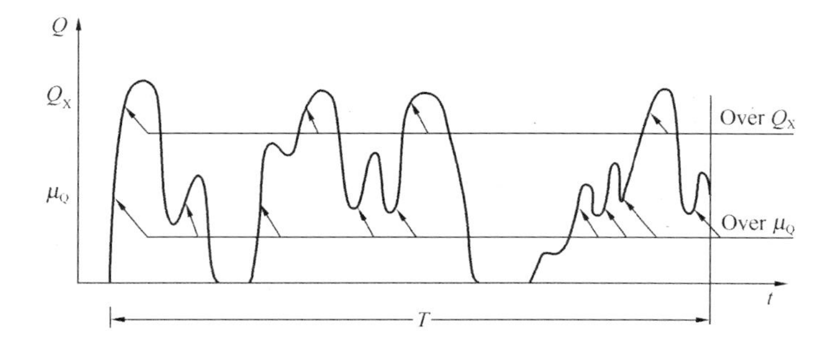

Figure C. 2. 2-2 Frequent Value of a Variable Action Defined with the crossing rate

The crossing rate may be determined through direct observations or computed by means of certain characteristics (for instance, spectral density function) of the stochastic process. If the mean μ_{Q^*} of action Q^* at arbitrary-time-point and crossing rate v_m are known, and the action is a stationary ergodic gauss stochastic process, then the action level Q_x corresponding to the crossing *rate* v_x may be calculated according to the following equation:

$$Q_x = \mu_{Q^*} + \sigma_{Q^*} \sqrt{\ln (\nu_m/\nu_x)^2} \qquad (C.2.2\text{-}4)$$

Where σ_{Q^*} ——standard deviation of action Q^* at arbitrary-time- point.

The frequent value, $\psi_f Q_k$, of action for the serviceability limit states related to the exceeding number of an action may be determined according to this method. The frequent value may be used to judge the serviceability of structure if the structural vibration involves the human comfortability, influences such limit states as the performance of non-structural elements and operation of equipment.

C. 2. 3 The quasi-permanent value of a variable action may be determined according to the following principle:

1 The average value of appearing part may be adopted as the quasi-permanent value $\psi_q Q_k$ for part of variable action that frequently appears on structure.

2 The quasi-permanent value of the variable action that is uneasy to be discriminated may be determined according to the specified ratio of the total duration when the value of action is exceeded to the design reference period. In this situation, the ratio may be taken as 0. 5. The quasi-permanent value $\psi_q Q_k$ may be determined directly according to equation (C. 2. 2-3) if the variable action may be considered as an ergodic stochastic process.

C. 2. 4 The combination value of a variable action may be determined according to the following principle:

1 The equal time interval model is adopted and assumed that the stochastic processes of all actions $Q(t)$ are stationary ergodic square wave process composed of equal time interval τ (Figure C. 2. 4).

2 The actions are ordered in sequence according to the number, r, of the time interval of each action within design reference period. Do the combination with one of all actions in the combination taking the maximal Q_{max} of the design reference period while the others taking the maximal of its own time interval or the value at arbitrary-time-point, which is generally named as combination value Q_c.

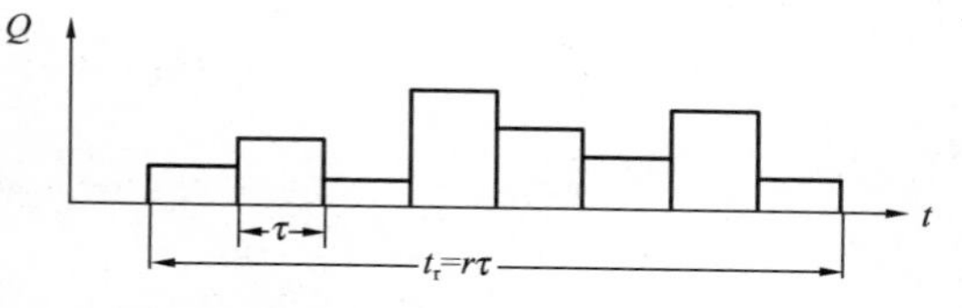

Figure C. 2. 4 Square Wave Stochastic Process with Equal Time Interval

3 The maximum design value, Q_{maxd}, of an action and the combination value Q_{cd} are determined as follows based on the design value method:

$$Q_{maxd} = F^{-1}_{Q_{max}}[\Phi(0.7\beta)] \qquad (C.2.4\text{-}1)$$

$$Q_{cd} = F^{-1}_{Q_c}[\Phi(0.28\beta)] \qquad (C.2.4\text{-}2)$$

$$\psi_c = \frac{Q_{cd}}{Q_{maxd}} = \frac{F^{-1}_{Q_c}[\Phi(0.28\beta)]}{F^{-1}_{Q_{max}}[\Phi(0.7\beta)]} = \frac{F^{-1}_{Q_{max}}[\Phi(0.28\beta)^r]}{F^{-1}_{Q_{max}}[\Phi(0.7\beta)]} \qquad (C.2.4\text{-}3)$$

The combination value factor may be presented as follows for the action following extreme value I-type distribution:

$$\psi_c = \frac{1 - 0.78v\{0.577 + \ln[-\ln(\Phi(0.28\beta))] + \ln r\}}{1 - 0.78v\{0.577 + \ln[-\ln(\Phi(0.7\beta))]\}} \tag{C.2.4-4}$$

Where v——variation coefficient of maximum value of an action.

4 The combination value factor may also be determined according to the method in Appendix E. 6.

Appendix D Test-Assisted Design

D.1 General Requirements

D.1.1 The test-assisted design shall comply with the following requirements:

1 The testing program shall be formulated before carrying out the test, including the necessary explanation on test objective, selection and manufacturing of specimens as well as the test implementation and assessment;

2 Qualitative analysis shall be carried out in advance to determine the possible critical region and corresponding limit state marks of the performance of the considered structure or structural member to formulate the testing program;

3 The specimens shall be manufactured by the same procedure as the member in construction;

4 The influence of sample size shall be taken into consideration in determination of the design value based on the test result.

D.1.2 The difference between the test condition and the physical condition of structure shall be considered through appropriate conversion or correction coefficient. The conversion coefficient η shall be determined through test or theoretical analysis. The major factors influencing the conversion coefficient η include size effect, time effect, and boundary condition of specimens, environmental conditions, and processing condition.

D.2 Statistical Assessment Principle of Test Result

D.2.1 The statistical assessment shall comply with the following fundamental principles:

1 The performance and failure mode of specimens shall be compared with the theoretical predicted value, and the reasons shall be explained and additional test shall be made if great difference exists.

2 The test results shall be assessed based on the available probability distribution and parameter information. The method specified in this appendix is only applicable to the case that the statistical data (or prior information) is from the same population;

3 The assessment results of the test is only valid for the considered test conditions and should not be extrapolated for other application.

D.2.2 The determination of design value of material property, model parameter or resistance shall comply with the following fundamental principles:

1 The design value of material property, model parameter or resistance may be derived by dividing the characteristic value, based on classical statistic method or "Bayesian method", with partial factor, and the influence of conversion coefficient may be considered if necessary;

2 The statistical uncertainty related to the sampling size and a priori statistical knowledge shall be taken into consideration in assessment of the design value of material property, model parameter or resistance, and the variability of test data.

D. 3 Statistical Assessment of the Design Value of Single Performance Index

D. 3. 1 The statistical assessment of the design value of single performance shall comply with the following general provisions:

1 The single performance X may refers to the resistance of member or some properties consisting of the resistance of the member;

2 All the conclusions in D. 3. 2 and D. 3. 3 are presented on the basis that the member resistance or the properties consisting of the member resistance following normal distribution or logarithmic normal distribution;

3 The design value generally may be estimated based on the classical approach if the prior knowledge on average value is unavailable, where "δ_x is unknown" is the case without prior knowledge of variation coefficient, "δ_x is known" is the case that the information of variation coefficient is known;

4 The design value may be estimated based on the Bayesian method if the prior knowledge of average value is available.

D. 3. 2 Classical statistical method

1 The design value X_d of the performance X may be expressed as follows if X follows normal distribution:

$$X_d = \eta_d \frac{X_{k(n)}}{\gamma_m} = \frac{\eta_d}{\gamma_m} \mu_x (1 - k_{nk} \delta_x) \quad \text{(D. 3. 2-1)}$$

Where η_d —design value of conversion coefficient, and the assessment of conversion coefficient depends mainly on the test type and materials;

γ_m —partial factor, its value shall be selected according to the application area of test result;

k_{nk} —one-sided tolerance coefficient;

μ_x —average value of performance X;

δ_x —variation coefficient of performance X.

2 Equation (D. 3. 2-1) may be rewritten as follows if the performance X follows logarithmic normal distribution:

$$X_d = \frac{\eta_d}{\gamma_m} \exp (\mu_y - k_{nk} \sigma_y) \quad \text{(D. 3. 2-2)}$$

Where μ_y —average value of variable $Y = \ln X$, and $\mu_y = m_y = \frac{1}{n} \sum_{i=1}^{n} \ln x_i$;

σ_y —standard deviation of the variable $Y = \ln X$;

$\sigma_y = \sqrt{\ln (\delta_x^2 + 1)}$ when δ_x is known;

$\sigma_y = S_y = \sqrt{\frac{1}{n-1} \sum_{i=1}^{n} (\ln x_i - m_y)^2}$ when δ_x is unknown;

x_i —i^{th} observed value of performance X.

D. 3. 3 Bayesian method

1 The design value of the performance X may be determined by the following equation if X follows the normal distribution:

$$X_d = \eta_d \frac{X_{K(n)}}{\gamma_m} = \frac{\eta_d}{\gamma_m} (m'' - k_{n\upsilon} \sigma'') \quad \text{(D. 3. 3-1)}$$

in which $k_{n\upsilon} = t_{p,\upsilon''}\sqrt{1+\frac{1}{n''}}$, $n'' = n' + n$,

$$\upsilon'' = \upsilon' + \upsilon + \delta(n'), m''n'' = m'n' + m_x n,$$

$$[(\sigma'')^2\upsilon'' + (m'')^2 n''] = [(\sigma')^2\upsilon' + (m')^2 n'] + [(\sigma_x)^2\upsilon + (m_x)^2 n]$$

Where $t_{p,\upsilon''}$ ——argument value of fractile p corresponding to the t-distribution with freedom of υ'', $P_t\{x > t_{p,\upsilon''}\} = p$;

$m', \sigma', n', \upsilon'$ ——prior distribution parameters.

2 The determination of prior distribution parameter n' and υ' shall comply with the following principles:

1) n' and υ' shall be taken as zero if the effective data is not sufficient. In this situation, the assessment results based on the Bayesian method and classical statistical method for the case of "δ_x is unknown" of are the same;

2) n' and υ' may be taken a larger value if the average value and standard deviation are considered to be a constant based on the past experience , such as 50 or the much larger one;

3) Generally, the value may be estimated by assuming that the data is insufficient or no prior information, and taking $n' = 0$ in this way.

Appendix E Basis for Structural Reliability Analysis and Design

E. 1 General Requirements

E. 1. 1 Analytical calculation and experience judgment should be combined in the determination of the partial factor and combination value factor according to the method specified in this Appendix, and adjustment may be required if necessary.

E. 1. 2 The following requirements for reliability analysis and design of structures according to this Appendix shall beas satisfied;

1 The structural limit state equation is available;

2 The statistical parameter and probability distribution of basic variables is accurate and reliable.

E. 1. 3 Action combination shall be conducted in the case of having two or more variable actions, and may be carried out by one of the following rules:

1 It is assumed that there are m's actions in the combination. Firstly, arrange the actions from smaller one to larger one according to the number, r_i, of the total time interval of modeled action $Q_i(t)$ within design reference period T, namely $r_1 \leqslant r_2 \leqslant \cdots \leqslant r_m$. Secondly, combine, in turn, the maximum value $\max\limits_{t \in [0,T]} Q_i(t)$ of any $Q_i(t)$ within (0, T) with the rest actions and get m's combinations with respect to maximal $Q_{\max,j}$ ($j = 1, 2, \cdots, m$). The maximum combination is the critical.

2 It is assumed that there are m's actions in the combination. Combine, in turn, the maximum $\max\limits_{t \in [0,T]} Q_i(t)$ of any action $Q_i(t)$ within (0, T) and $Q_j(t_0)$ ($i \neq j$) of the rest action at arbitrary-time- point and get m's action combinations with respect to maximal $Q_{\max,j}$ ($j = 1, 2, \cdots, m$). The maximum combination is the critical.

E. 2 Calculation of Structural Reliability Index

E. 2. 1 The reliability index of structure or its member should be calculated by the first-order reliability method considering the probability distribution of random variables, or other methods.

E. 2. 2 The following requirements shall be complied with if the first-order reliability method is adopted:

1 The reliability index of structure or structural member may be calculated the following equation if only two mutual independent integrated varibles, action effect and structural resistance, are involved and both follow normal distribution:

$$\beta = \frac{\mu_R - \mu_S}{\sqrt{\sigma_R^2 + \sigma_S^2}} \quad \text{(E. 2. 2-1)}$$

Where β——reliability index of structure or structural member;

μ_S, σ_S——mean and standard deviation of the action effect of structure or structural member;

μ_R, σ_R——mean and standard deviation of the resistance of structure or structural member.

2 The reliability index of structure or structural member shall be iteratively calculated according to the following equations if there are n's mutual independent non-normal basic variables based on the limit state equation (4.3.5):

$$\beta=\frac{g(x_1^*,x_2^*,\cdots,x_n^*)+\sum_{j=1}^{n}\frac{\partial g}{\partial X_j}\bigg|_P(\mu_{X'_j}-x_j^*)}{\sqrt{\sum_{j=1}^{n}\left(\frac{\partial g}{\partial X_j}\bigg|_P\sigma_{X'_j}\right)^2}} \tag{E.2.2-2}$$

$$\alpha_{X'_i}=-\frac{\frac{\partial g}{\partial X_i}\bigg|_P\sigma_{X'_i}}{\sqrt{\sum_{j=1}^{n}\left(\frac{\partial g}{\partial X_j}\bigg|_P\sigma_{X'_j}\right)^2}}(i=1,2,\cdots,n) \tag{E.2.2-3}$$

$$x_i^*=\mu_{X'_i}+\beta\alpha_{X'_i}\sigma_{X'_i}(i=1,2,\cdots,n) \tag{E.2.2-4}$$

$$\mu_{X'_i}=x_i^*-\Phi^{-1}[F_{X_i}(x_i^*)]\sigma_{X'_i}(i=1,2,\cdots,n) \tag{E.2.2-5}$$

$$\sigma_{X'_i}=\frac{\varphi\{\Phi^{-1}[F_{X_i}(x_i^*)]\}}{f_{X_i}(x_i^*)}(i=1,2,\cdots,n) \tag{E.2.2-6}$$

Where $g(\cdot)$——performance function of structure or member;

$X_i(i=1,2,\cdots,n)$——basic variable;

$x_i^*(i=1,2,\cdots,n)$——coordinate value of basic variable X_i at the design point;

$\frac{\partial g}{\partial X_i}\Big|_P$——value of the first order partial derivative of performance function g $(X_1, X_2,\cdots, X_n)$ at design point;

$\mu_{X'_i}$、$\sigma_{X'_i}$——mean and standard deviation of the equivalent normalized variable X'_i of basic variable X_i;

$f_{X_i}(\cdot)$, $F_{X_i}(\cdot)$——probability density function and probability distribution function of basic variable X_i;

$\varphi(\cdot)$, $\Phi(\cdot)$, $\Phi^{-1}(\cdot)$——probability density function, probability distribution function and inverse function of probability distribution function of standard normal random variable.

3 The iterative calculation may carried out after replacing the equation (E.2.2-2) and Equation (E.2.2-3) with the following equations respectively if some non-normal and correlated basic variables are present based on the limit state equation (4.3.5):

$$\beta=\frac{g(x_1^*,x_2^*,\cdots,x_n^*)+\sum_{j=1}^{n}\frac{\partial g}{\partial X_j}\bigg|_P(\mu_{X'_j}-x_j^*)}{\sqrt{\sum_{k=1}^{n}\sum_{j=1}^{n}\left(\frac{\partial g}{\partial X_k}\bigg|_P\frac{\partial g}{\partial X_j}\bigg|_P\rho_{X'_k,X'_j}\sigma_{X'_k}\sigma_{X'_j}\right.}} \tag{E.2.2-7}$$

$$\alpha_{X'_i}=-\frac{\sum_{j=1}^{n}\frac{\partial g}{\partial X_j}\bigg|_P\rho_{X'_i,X'_j}\sigma_{X'_j}}{\sqrt{\sum_{k=1}^{n}\sum_{j=1}^{n}\frac{\partial g}{\partial X_k}\bigg|_P\frac{\partial g}{\partial X_j}\bigg|_P\rho_{X'_k,X'_j}\sigma_{X'_k}\sigma_{X'_j}}}\quad(i=1,2,\cdots n) \tag{E.2.2-8}$$

Where $\rho_{X'_i,X'_j}$——correlation coefficient of equivalent normalized variables X'_i and X'_j, which may be approximately taken as the correlation coefficient ρ_{X_i,X_j} of variables X_i and X_j.

E.3 Calibration of Structural Reliability

E.3.1 The calibration of structural reliability is actually a process to analyze the reliability level of the structure designed according to the traditional method, and is the basis to determine the reliability index to be adopted in the design of the sturcutres. The structure or structural member selected for calibration shall be typical and representative.

E. 3. 2 The following steps may be followed for structural reliability calibration:

1 Determining the calibration range, such as selecting the structure type (for example, building structure, bridge structure, and part and harbor structure) or the construction material (such as concrete structure and steel structure), and selecting the representative structures or structural members (including the failure mode of member) according to the application scope of the target reliability index;

2 Determining the range of basic variables in the design, such as the range of the ratio between the characteristic value of a variable action and the characteristic value of a permanent action;

3 Analyzing the format of traditional design method, such as the expressions for flexural and shear design;

4 Calculating the reliability index β_i of different structure or structural member;

5 Determining a sets of weights ω_i, which shall comply with equation (E. 3. 2-1), according to the scope of application and importance of structure or structural member in practice:

$$\sum_{i=1}^{n} \omega_i = 1 \tag{E. 3. 2-1}$$

6 Determining the weighted average of the reliability index of calibrated structure or structural member according to the following equation:

$$\beta_{\text{ave}} = \sum_{i=1}^{n} \omega_i \beta_i \tag{E. 3. 2-2}$$

E. 3. 3 The target reliability index β_t of structure or structural member shall be determined based on the calibrated reliability β_{ave} combined with the comprehensive analysis and judgment.

E. 4 Design Based on Reliability Index

E. 4. 1 One of the following methods may be adopted for the design of structure or structural member based on the target reliability index:

1 The following expression shall comply with for the reliability index of the designed structure or structural member:

$$\beta \geqslant \beta_t \tag{E. 4. 1-1}$$

Where β——reliability index of the designed structure or structural member;

β_t——target reliability index of designed structure or structural member.

The structure or structural member shall be redesigned if the expression (E. 4. 1-1) is not satisfied.

2 The geometrical parameter related to the load carrying capacity of the structural member may be directly quantitated according to the following equation which meet the requirement of expression (E. 4. 1-1) for the cross-section design of certain structural members, such as the amount of the reinforcement at cross-section of reinforced concrete member, in the case of the resistance follows logarithmic normal distribution:

$$\frac{R(f_k, a_k)}{k_R} = \sqrt{1+\delta_R^2} \exp\left(\frac{\mu_{R'}}{r^*} - 1 + \ln r^*\right) \tag{E. 4. 1-2}$$

Where $R(\cdot)$——resistance function;

$\mu_{R'}$—— mean of normalized resistance calculated iteratively;

r^*——design point value of resistance calculated iteratively;

δ_R——variation coefficient of resistance;

f_k——characteristic value of material property;

a_k——characteristic value of a geometrical parameter, such as the cross-sectional area of steel reinforcement in reinforced concrete member;

k_R——bias or the ratio between mean and characteristic value of resistance.

E. 4. 2 The reasons for the difference shall be analyzed carefully in case that there is significant difference between the design results obtained by reliability index method and the traditional method. The design based on the reliability index method shall not be adopted until it is proved to be reasonable.

E. 5 Determination of Partial Factors

E. 5. 1 The following principles shall comply with for the determination of the partial factors in the design expression of structure or structural member:

1 The same action partial factors shall be adopted for the same actions on the structure , and different action partial factors for different actions;

2 Different resistance partial factors shall be adopted for different kinds of members , and the resistance partial factor of member shall be unchanged under any variable actions;

3 The reliability indexes of the members of different kinds designed according to the selected action partial factors and resistance factor shall be in the best consistent with the target reliability index β_t under different ratio of action effects.

E. 5. 2 The partial factors in the design expression of structure or structural member may be determined as the following steps:

1 Select a representative structure or structural member (or failure mode), one simple combination composed of one permanent action and one variable action (for example, permanent action + variable action on floor on building structure, or permanent action + wind action) and typical action effects ratio (ratio between the characteristic values of the variable action effect and the permanent action effect);

2 The importance factor γ_0 shall be 1. 0 for the structure or structural member with safety class Ⅱ;

3 Determine the design value of resistance for the given partial factor γ_G and γ_Q for the selected structure or structural members;

4 Determining the characteristic value of resistance of the given resistance factor γ_R for the selected structure or structural member;

5 Calculate the reliability index β in the simple combination for the selected structure or structural member;

6 Determine the γ_R optimally for the all selected representative structures or structural members and in the selected range of γ_G and γ_Q (with differential of 0. 1 or 0. 05); select a set of partial factors γ_G, γ_Q and γ_R that make the reliability index β of structure or structural member designed according to partial factor expression to be in the best consistent with the target reliability index β_t;

7 Examine the set of partial factor γ_G, γ_Q and γ_R based on the engineering experience, and making an adjustment if necessary;

8 The partial factor of permanent action takes a minus sign in the design expression if the permanent action is favorable, and the partial factor γ_G shall be determined optimally (in differential of 0.1 or 0.05) based on the selected partial factor γ_Q and γ_R;

9 Determine the importance factor γ_0 for the structure or structural member with safety class I and III based on the set of partial factors of the structure or structural member for safety class II, and make the reliability index β of structure or structural member designed according to partial factor expression being in the best consistent with the target reliability index β_t.

E.6 Determination of Combination Value Factor

E.6.1 The determination of the combination value factor of a variable action shall comply with the following principles:

The determined combination value factor shall make the reliability index β of structures or structural members designed according to partial factor expression being in the best consistent with the target reliability index β_t in the case with two or more than two variable actions participating in combination and the variable action partial factor γ_G, γ_Q and the resistance partial factor γ_R being determine.

E.6.2 The combination value factor of a variable action may be determined follows the following steps:

1 The determination of combination value factor of a variable action shall be based on the structure or structural member with safety class II. Select the representative structure or structural member (or failure mode), as well as the combination composed of one permanent action and two or more variable actions and the typical action effects ratio (the ratio of the leading characteristic values of variable action effect to permanent action effect, and the ratio of the accompanying characteristic values of variable action effect to the leading characteristic value of variable action effect);

2 Calculate the resistance design value for selected structures or structural members, different load combination and typical action effect ratios based on the determined partial factor γ_G and γ_Q;

3 Calculate the resistance characteristic value for selected structures or structural members, different load combination and typical action effect ratios based on the determined resistance partial factor γ_R;

4 Calculate the reliability index of for selected structures or structural members and different load combinations for the typical action effect ratio;

5 Determine the combination value factor ψ_c for all the selected representative structures or structural members, load combinations and typical action effect ratios to make the reliability index β of structures or structural members designed according to the partial factor expression being in the best consistent with the target reliability index β_t;

6 Examine the determined combination value factor ψ_c by engineering judgment, and make an adjustment it if necessary.

Appendix F Verification of Fatigue Reliability of Structure

F.1 General Requirements

F.1.1 This Appendix is applicable to verify the fatigue reliability of engineering structures, including building structure, railway and highway bridge and culvert structure, and municipal engineering structure, subjected to high cycle fatigue action.

F.1.2 The fatigue reliability of structure or construction shall be verified at the following conditions:

1 The entire or parts of the structure subjected to repeat loading;

2 The structure or its detailing work with remarkable stress concentration and undergone alternative stress;

3 The duration time of repeat loading accounts for the main portion of the design working life of structure.

F.1.3 The fatigue reliability of structures may be verified respectively for the ultimate limit states or serviceability limit states as required.

F.1.4 The fatigue reliability of one or more detailing of the structure may be verified respectively.

F.1.5 The fatigue reliability of structure shall be verified by the following steps:

1 Determine the key parts and detailing of the structure or the parts assured by the client to be verified according to the structural analysis results ;

2 Formulate standard load fatigue spectrum according to the investigation of the load history that the structure had experienced in service period and forecast;

3 Analyze and assess the fatigue action on the structure or parts of the structure and the corresponding fatigue resistance;

4 Draw a conclusion of verification of the fatigue reliability.

F.1.6 The mechanical model and internal force calculation method involved in this Appendix shall comply with the relevant regulations specified in Chapter 7.

F.1.7 The criterion for fatigue capacity verification of structure shall be that the calculated nominal stress at the checked position does not exceed the design value of fatigue strength at the corresponding position of structure.

F.1.8 The design value of fatigue strength shall be expressed in nominal stress(at position where stress does not concentrate), and shall be defined as a certain upper fractile of probability distribution based on fatigue test results of structure or parts of structur .

F.1.9 The determination of the target reliability index in the fatigue verification may be based on fatigue calibration.

F.2 Fatigue Action

F.2.1 The process of the variable amplitude repeated load sustained by the structure may be determined by measurement or simulation. The loading process may be converted into load

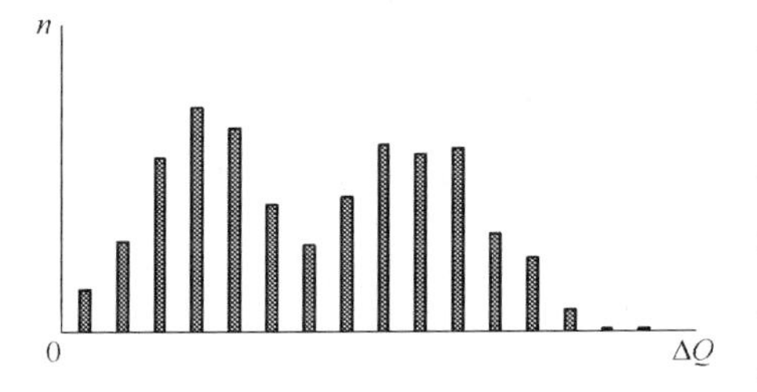

Figure F. 2. 1 Load Spectrum

spectrum (Figure F. 2. 1) representing the relation between the load range ΔQ ($\Delta Q = Q_{max} - Q_{min}$) and number of load cycles by "rain-flow counting method" or "reservoir method". The "load spectrum" may be converted into stress spectrum of the structure, connection or local parts of structure. In which, the stress range $\Delta\sigma = \sigma_{max} - \sigma_{min}$ may be determined through the load range ΔQ.

F. 2. 2 The stress spectrum of structural member (or connection) can be converted into the equivalent constant-amplitude repeated stress with specified number of cycles based on the "Miner damage accumulation criterion", and then formed equivalent fatigue action taking the necessary influence factor into consideration (including dead load if necessary). Generally, the specified number of load cycles of equivalent constant-amplitude repeated stress may be 2×10^6.

The calculation for the fatigue action of details of steel structure and concrete structure is as follows:

1 Fatigue action of steel structure

The equivalent fatigue action of steel structure may be calculated according to the following equation.

$$\Delta\sigma_{aek} = K_{a1}K_{a2}K_{a3}\cdots K_{ai}\Delta\sigma_{ac} = (\prod_{i=1}^{m} K_{ai})\Delta\sigma_{ac} \quad \text{(F. 2. 2-1)}$$

Where $\Delta\sigma_{aek}$——characteristic value of the equivalent fatigue stress range at the checking position of steel structure;

$\Delta\sigma_{ac}$——characteristic value of the stress range at verification position of steel structure with action of characteristic value of load;

K_{ai}——i^{th} fatigue influence factor of steel structure whose value shall be determined by the ratio between the statistical result under corresponding influence and $\Delta\sigma_{ac}$, and it shall be coordinated with the $\Delta\sigma_{ac}$ and the specified number of cycles of the characteristic value of corresponding fatigue resistance;

m——number of the fatigue influence factors related to the structure.

2 Fatigue action of concrete structure

The equivalent fatigue actions of concrete structure may be calculated according to the following equations.

$$\sigma_{cek} = K_{c1}K_{c2}K_{c3}\cdots K_{ci}\sigma_{cc} = (\prod_{i=1}^{n} K_{ci})\sigma_{cc} \quad \text{(F. 2. 2-2)}$$

$$\Delta\sigma_{pek} = K_{p1}K_{p2}K_{p3}\cdots K_{pi}\Delta\sigma_{pc} = (\prod_{i=1}^{n} K_{pi})\Delta\sigma_{pc} \quad \text{(F. 2. 2-3)}$$

$$\Delta\sigma_{sek} = K_{s1}K_{s2}K_{s3}\cdots K_{si}\Delta\sigma_{sc} = (\prod_{i=1}^{n} K_{si})\Delta\sigma_{sc} \quad \text{(F. 2. 2-4)}$$

Where σ_{cek}、$\Delta\sigma_{pek}$、$\Delta\sigma_{sek}$——characteristic value of the equivalent concrete fatigue stress, the characteristic value of fatigue stress range of the equivalent prestressed reinforcement and nonprestressed reinforcement at the checking position of concrete structure spectively;

σ_{cc}、$\Delta\sigma_{pc}$、$\Delta\sigma_{sc}$——characteristic value of the concrete stress, the characteristic value of stress range of the prestressed reinforcement and the nonprestressed reinforcement at the checking position of concrete structure under the action of characteristic value of load spectively;

K_{ci}、K_{pi}、K_{si} —— i^{th} fatigue influencing parameter of concrete, prestressed reinforcement and nonprestressed reinforcement at the checking position of concrete structure. Their values are separately determined by the ratio between the statistical result of corresponding influence factors and the corresponding σ_{cc}, $\Delta\sigma_{pc}$, and $\Delta\sigma_{sc}$, and shall be coordinated with the numbers of cycles specified by the σ_{cc}, $\Delta\sigma_{pc}$, $\Delta\sigma_{sc}$ as well as corresponding characteristic value of fatigue resistance respectively;

n —— number of influencing factors concrete sturcture, which is related to the structural pattern.

F. 2. 3 The probability distribution and the statistical parameters of influencing factors in the fatigue actions may be determined by statistical method and its characteristic value shall be the same average value as the static force action.

F. 3 Fatigue Resistance

F. 3. 1 The fatigue resistance refers to the capability of structure or parts of structure resisting the fatigue action of specified number of cycles.

F. 3. 2 The fatigue resistance of material and non-welded steel structure are related to the maximum stress σ_{max} resulted from the fatigue action and the stress ratio ρ as well as the detailing of structure. The fatigue resistance of welded steel structure is related to the stress range $\Delta\sigma$ resulted from the fatigue action and the detailing of the structure. The calculation for the fatigue resistance of the detailing of steel structure and concrete structure are as follows:

1 Fatigue resistance of steel structure

The expression of the fatigue resistance of steel structure may be expressed by the $S-N$ fatigue equation shown in Equation (F. 3. 2-1);

$$\Delta\sigma^{m}N = C \tag{F. 3. 2-1}$$

Where $\Delta\sigma$ —— constant-amplitude fatigue stress range of the detailing at checking position of steel structure ,MPa;

N —— number of cycles to failure;

m, C —— fatigue parameters determined through fatigue test.

The fatigue resistance Δf_{aek} of steel structural member refers to the characteristic value of the maximal fatigue stress range determined by Equation (F. 3. 2-1) under the specified number of cycles and specified insurance probability at the checking position of steel structure.

2 Fatigue resistance of concrete structure

1) Concrete

The factors influencing the concrete fatigue resistance for concrete structure include the fatigue strength, fatigue elastic modulus and fatigue deformation modulus.

The characteristic value of the fatigue strength of concrete may be determined by multiplying the characteristic value of the static strength of concrete with an equivalent reduction factor:

$$f_{cek} = K_{ce} f_{ck} \quad (F.3.2\text{-}2)$$

Where f_{cek}——characteristic value of concrete fatigue strength;

K_{ce}——reduction factor of concrete fatigue strength related to such factor as minimum value of the concrete stress;

f_{ck}——characteristic value of the concrete static strength.

The fatigue elastic modulus of concrete may be determined by test. For the under-reinforced concrete bending member, the characteristic value of the fatigue elastic modulus of fatigue may be the characteristic value of static elastic modulus being multiplied by 0.7.

The fatigue deformation modulus of concrete may be determined by test. For the under-reinforced concrete bending member, the characteristic value of the fatigue deformation modulus of concrete may be the characteristic value of static deformation modulus being multiplied by 0.6.

2) Prestressed or ordinary reinforcement

The fatigue strength of prestressed or ordinary reinforcement in concrete structure may be determined through the $S-N$ fatigue equation showed in Equation (F.3.2-1). The fatigue resistance Δf_{pek} or Δf_{sek} refers to the characteristic value of the maximaum fatigue stress range of prestressed or ordinary reinforcement determined by Equation (F.3.2-1) under specified number of cycles and specified insurance probability at the checking position of concrete structure.

F.4 Verification of Fatigue Reliability

F.4.1 Generally, the fatigue reliability of steel structure shall be verified with respect to the fatigue ultimate limit states and may be conducted by the equivalent constant amplitude repeated stress-based method, limit damage tolerance-based method and fracture mechanics-based method as required.

1 Equivalent constant-amplitude repeated stress-based method

1) The following requirement shall be satisfied for the verification if the equivalent constant-amplitude repeated stress-based method is expressed in working stress format:

$$\Delta\sigma_{aek} \leqslant \Delta f_{aek} \quad (F.4.1\text{-}1)$$

2) The design value of a fatigue action may be the characteristic value of the equivalent constant-amplitude repeating action of the nominal fatigue load effect in the design working life of structural member being multiplied by the partial factor of fatigue action if the equivalent constant-amplitude repeated stress-based method is expressed in partial factor format. The fatigue resistance may be determined through the fatigue test with number of cycles the same with the equivalent constant-amplitude repeating action. The following requirement shall be complied with for fatigue verification:

$$\gamma_0 \gamma_{aek} \Delta\sigma_{aek} \leqslant \frac{\Delta f_{aek}}{\gamma_{af}} \quad (F.4.1\text{-}2)$$

Where γ_0——Importance factor of structure;

γ_{aek}——Partial factor considering the uncertainty of equivalent constant-amplitude fatigue action and fatigue action model;

γ_{af}——Partial factor of fatigue resistance, and $\gamma_{af}=1.0$ when the insurance probability for the value of fatigue resistance is 97.7%.

2 Limit damage tolerance-based method

1) The fatigue damage tolerance-based verification shall comply with the Palmgren-Miner linear cumulative damage rule as expressed in Equation (F. 4. 1-3), that is, the sum of the ratio between the number of cycles under the stress range level $\Delta\sigma_i$ and the number of cycles under the same stress range level to failure shall less than the fatigue damage tolerance:

$$\Sigma \frac{n_i}{N_i} < D_c \tag{F. 4. 1-3}$$

Where n_i——number of cycles of fatigue actions the structure subjected to corresponding to the stress range level $\Delta\sigma_i$. If the range level, $\Delta\sigma_i$, of fatigue stress is less than a certain value, $\Delta\sigma_0$, the corresponding cycles of fatigue action shall be reduced by multiplying with $\left(\frac{\Delta\sigma_i}{\Delta\sigma_0}\right)^2$;

N_i——number of cycles to failure under stress range level $\Delta\sigma_i$;

D_c——Critical fatigue damage tolerance, and it is 1.0 at perfect.

2) The following requirement shall be complied with for the fatigue verification if the limit damage tolerance-based method is expressed in partial factor format:

$$\Sigma \frac{n_i}{N_i} < \frac{D_c}{\gamma_d} \tag{F. 4. 1-4}$$

$$N_i = N_i\left(\gamma_d, \gamma_{\Delta\sigma_i} \Delta\sigma_i, \frac{\Delta f_{aek}}{\gamma_{ak}}\right) \tag{F. 4. 1-5}$$

Where γ_d——partial factor related to the cumulative damage criteria, design working life and the uncertainty of failure consequence;

$\gamma_{\Delta\sigma_i}$——partial factor considering the uncertainty of the fatigue stress range level and fatigue action model;

γ_{ak}——partial factor considering the uncertainty of fatigue resistance model of materials and detailing.

3 Fracture mechanics-based method

The fracture mechanics-based verification method shall be adopted when the steel structure operates at low temperature environment.

F. 4. 2 The fatigue of concrete and steel reinforcement shall be verified respectively for the fatigue ultimate limit states verification of concrete structures. The equivalent constant-amplitude repeated stress-based method and limit damage tolerance-based method may be adopted as required.

1 Equivalent constant-amplitude repeated stress-based method

1) The following requirement for the fatigue verification of the concrete and prestressed reinforcement at the checking position of structure shall be complied with if the equivalent constant-amplitude repeated stress-based method is expressed in working stress format:

$$\sigma_{cek} \leqslant f_{cek} \tag{F. 4. 2-1}$$

$$\Delta\sigma_{pek} \leqslant \Delta f_{pek} \tag{F. 4. 2-2}$$

$$\Delta\sigma_{sek} \leqslant \Delta f_{sek} \tag{F. 4. 2-3}$$

2) The design value of fatigue action may be the characteristic value of the equivalent constant-amplitude repeating action of the nominal fatigue load effect in the design working life of structural member multiplying the partial factor of fatigue action if the

equivalent constant-amplitude repeated stress-based method is expressed in partial factor format. The fatigue resistance may be determined through fatigue test with number of cycles the same as the equivalent constant-amplitude repeating action. The following requirement for fatigue verification of concrete, prestressed and ordinary reinforcement at checking position of structure shall be complied with:

$$\gamma_0 \gamma_{cek} \sigma_{cek} \leqslant \frac{f_{cek}}{\gamma_{cf}} \tag{F. 4. 2-4}$$

$$\gamma_0 \gamma_{pek} \Delta\sigma_{pek} \leqslant \frac{\Delta f_{pek}}{\gamma_{pf}} \tag{F. 4. 2-5}$$

$$\gamma_0 \gamma_{sek} \Delta\sigma_{sek} \leqslant \frac{\Delta f_{sek}}{\gamma_{sf}} \tag{F. 4. 2-6}$$

Where γ_{cek}, γ_{pek}, γ_{sek}——partial factors considering the uncertainty of fatigue action model of equivalent constant-amplitude fatigue action of concrete, prestressed and ordinary reinforcement respectively;

γ_{cf}, γ_{pf}, γ_{sf}——partial factors of the fatigue resistance of concrete, prestressed and ordinary reinforcement respectively.

2 Limit damage tolerance-based method

The limit damage tolerance-based reliability verification method of concrete structure with respect to fatigue ultimate limit states is the same as the fatigue verification method of steel structure specified in the Item 2 of Article F. 4. 1 in this Appendix, in which the materials at the checking positions shall be concrete, prestressed and ordinary reinforcement.

F. 4. 3 The serviceability limit states equation shall be established if the fatigue reliability verification with respect to serviceability limit state of structure is required, and the fatigue action effect may be superposed linearly. The increase in deformation caused by fatigue of material shall be taken into consideration in the calculation of limit value of fatigue.

Appendix G Reliability Assessment of Existing Structures

G. 1 General Requirements

G. 1. 1 This Appendix is applicable to the reliability assessment of existing structures designed and constructed according to relavant standards.

G. 1. 2 The reliability of existing structures should be assessed in the following conditions:

1 The working life of structure exceeds the design working life;

2 The usage or operating requirements of structure are changed;

3 The operating environment of structure gets severe;

4 The structure has some severe defects;

5 The structure are in the condition of material property deterioration, member damaging or other unflavored states that may influence its safety, serviceability and durability;

6 The reliability of existing structure is doubted or objected.

G. 1. 3 The reliability assessment of existing structure shall reduce the treatment for the structure possibly with precondition ensuring the structural performance.

G. 1. 4 The reliability assessment of existing structure include safety assessment, serviceability assessment and durability assessment, and the assessment on disaster resistance capability shall also be carried out if necessary.

G. 1. 5 The reliability assessment of existing structure shall be carried out according to specifications in current national related standards.

G. 1. 6 The reliability assessment of existing structure shall be made as the following steps:

1 Define the assessing objects, contents, and objectives;

2 Acquire the data and information related to the action on structure and the actual performance and conditions of the structure through survey or detection;

3 Analyze the reliability of the structure;

4 Work out the assessment report.

G. 2 Safety Assessment

G. 2. 1 The safety assessment of existing structure shall include three assessment respects as structural system and members arrangement, connection and detailing, and load carrying capacity.

G. 2. 2 The arrangement of the structural system and member of existing structure shall be assessed based on the specifications in current structural design standards.

G. 2. 3 The connection of existing structure and the detailing related to safety shall be assessed based on the specifications in current structural design standards.

G. 2. 4 The load carrying capacity may be assessed by the following methods according to the different conditions for the structure that the assessment result of the structural system and member arrangement as well as connection and detailing comply with the specification in Article G. 2. 2 and Article G. 2. 3:

1 Evaluation method based on good condition of the structure;

2 Evaluation method based on partial factor or lumped safety factor;

3 Evaluation method by adjusting the resistance partial factor based on reliability index;

4 Evaluation method based on load testing;

5 Other applicable evaluation methods.

G. 2. 5 The evaluation method based on good condition of structure should be adopted if the structure performs well. In this situation, the structure may be assessed to be safe if its load carrying capacity meets the following requirements:

1 No obvious problems related to the serviceability, such as deformation, crack, displacement, and vibration, are detected;

2 No significant changes for the action on structure and the environment in the assessed working life.

G. 2. 6 The structure is assessed to be safe if the following requirements are satisfied when the assessment is based on partial factor or lumped factor method:

1 The load carrying capacity of member shall be determined according to the structural computation models in the current structural design standard and the parameters in the models shall be adjusted based on actual condition:

1) The value of the material strength of member should be based on the measured data and shall be determined according to the method specified in the structural detection standards;

2) The geometrical parameters in computation models may be determined according to the actual size of members;

3) The adverse effect of irrecoverable damages shall be taken into consideration in calculation and analysis of the load carrying capacity of members;

4) The favorable influence to load carrying capacity of member may be considered in the computation module if it is verified

2 The action and action effect shall be determined according to the specification in the national current standards, and the following parameters or analysis methods may be adjusted:

1) The permanent action shall be determined according to the method specified in the current load standards of structure based on the investigation in site;

2) Some of the variable actions may be determined using the load adjustment factor considering design working life of structure based on the assessed working life;

3) The adverse effects of axial deviation, dimensional deviation and installation deviation shall be taken into consideration in the calculation of the action effect;

4) The action effect shall be determined based on the possible worst load combination.

3 The load carrying capacity of member determined based on the above methods shall not be less than the action effect or the lumped safety factor shall not be less that specified in the related standard for structural design.

G. 2. 7 The assessment method adjusting the resistance partial factor based on the reliability index may be used if the load carrying capacity and variation coefficient of one batch of members may be determined. In this situation, the batch of the members may be assessed to be safe if the following requirements are satisfied at the same time:

1 The calculation of action effect comply with the specification in Article G. 2. 6;

2 The resistance partial factor is adjusted according to the measured variation coefficient of the load carrying capacity of structural member;

3 The load carrying capacity calculated according to aforementioned principle does not less than the action effect.

G.2.8 Load testing-based method may be used for the assessment of structure or structural member if it is possible. The structure or structural member that meet the following requirements at the same time may be assessed to be safe:

1 The pattern of proof load shall be fundamentally accorded with the primary load on the structure, and the proof load shall not cause irreversible deformation or damage to the structure or member;

2 The results of load testing and the computational analysis shall comply with the specification in related standards.

G.2.9 Proposal for strengthening shall be provided for structure or structural member assessed to be dissatisfy the load carrying capacity requirements, and restriction for utilization may also be suggested if necessary.

G.3 Serviceability Assessment

G.3.1 The serviceability, which is related to deformation, crack, displacement and vibration, of structure shall be assessed based on the criteria specified in the current standard on structural design in the case of safety of structure being assured. The limit of serviceability limit states may be adjusted or determined according to the situations in site at the following conditions:

1 Obvious serviceability problems have been occur but has not reached the limit of serviceability limit states of the structure or member;

2 The criteria in relevant standards may be not reasonable for the verification of the serviceability of structure.

G.3.2 Suggestion for treatment shall be provided for the structure or member that the limit of serviceability limit states is exceeded.

G.3.3 Serviceability limit states should be assessed within the assessed working life for the structure or member that has not reached the limit . And the following principles should be abided by:

1 The computation model in the current standards for structural design may be used in the assessment, but the parameters of the model shall be calibrated based on investigation in site;

2 Load testing-based or field testing-based method may be used if it is possible;

3 Suggestion for treatment shall be provided if the structure or member is assessed to be unable to fulfill the serviceability requirements.

G.4 Durability Assessment

G.4.1 The durability assessment for the existing structures shall be carried out for the purpose of examining the relation between the durable years and assessed working life of the assessed structures.

Note: Durable years refers to the number of years the structures need to reach the limit of serviceability limit states under the environmental action.

G. 4. 2 The limit or sign of structures for serviceability limit states under the environmental action shall be determined according to the following principles:

1 No obvious surface damages impairing the load carrying capacity of the structural members occurs;

2 The deterioration of material properties increase the risk of brittle failure of the structural members.

G. 4. 3 The structural members served in the same environmental action and with the same material properties shall be taken as one batch in the determination of durable years of the existing structures.

G. 4. 4 The durable years of the batch structural members may be assessed based on the service period of the structure, changes of the material properties, environmental action and deterioration of the material properties.

G. 4. 5 Suggestions for maintenance shall be proposed for the structural members whose durable years are less than the assessed working life.

G. 5 Disaster Resistance Capacity Assessment

G. 5. 1 Disaster resistance capacity of the existing structures should be comprehensively assessed with respects to arrangement, connection, detailing, load carrying capacity, disaster prevention/reduction and protective measures of the structural system and members.

G. 5. 2 The disaster capacities for seismic, typhoon, sleet and flood as well as other natural disasters should be assessed by verifying the safety of structure .

G. 5. 3 The disaster resistance capacities for explosion, impact, fire and other accidental actions exerted on the locals of structures shall be assessed by evaluating measures for reducing the accidental action and its effect, preventing the structures from failure that disproportionate to the original cause, and measures for diminishing the scope of influence of accidental actions.

Measures for reducing the accidental action and its effect include explosion prevention and explosion-venting, collision and impact prevention , as well as controlling and fire- fighting facilities for the combustible matters.

Measures for reducing the scope of influence of the accidental actions include structural deformation joint installation and secondary hazard prevention.

G. 5. 4 The pre-warning and evacuation measures shall be assessed for the disasters that the structure may not resist.

Explanation of Wording in This Standard

1 Words used for different degrees of strictness are explained as follows in order to mark the differences in executing the requirements in this standard:

1) Words denoting a very strict or mandatory requirement:
"Must" is used for affirmation; "must not" for negation.

2) Words denoting a strict requirement under normal conditions:
"Shall" is used for affirmation; "shall not" for negation.

3) Words denoting a permission of a slight choice or an indication of the most suitable choice when conditions permit:
"Should" is used for affirmation; "should not" for negation.
"May" is used to express the option available, sometimes with the conditional permit.

2 "Shall comply with..." or "shall meet the requirements of..." is used in this code to indicate that it is necessary to comply with the requirements stipulated in other relative standards and codes.